AMERICAN WOODWORKING TOOLS

Croze-howel. Combination tool, with croze on one side and howel opposite. The arc of the working surfaces can be changed by adjustment of two lignum vitae wheels. Length, 12⅛ inches.

AMERICAN
WOODWORKING TOOLS

TEXT BY
PAUL B. KEBABIAN

PHOTOGRAPHS BY
DUDLEY WITNEY

NEW YORK GRAPHIC SOCIETY   BOSTON

First edition

LIBRARY OF CONGRESS CATALOGING IN PUBLICATION DATA

Kebabian, Paul B
    American woodworking tools.

    Bibliography: p.
    Includes index.
    1. Woodworking tools.   I. Witney, Dudley.   II. Title.
TT186.K4   621.9′08   78–7066
ISBN 0–8212–0731–8

Line drawings are by Cathy Baker, Hinesburg, Vermont

Designed by Janis Capone

New York Graphic Society Books are published by
Little, Brown and Company
Published simultaneously in Canada by Little, Brown
and Company (Canada) Limited
Printed in the United States of America

FOR JUSTINE AND PAMELA

CONTENTS

ACKNOWLEDGMENTS

I wish to express my gratitude to the following: my wife, Justine, for her assistance in preparation of the manuscript; Betty Childs, of New York Graphic Society, whose editorial expertise was essential; my brother, John S. Kebabian, who introduced me to the collecting and study of early American woodworking tools, and whose contributions to the history of American tools and manufacturing firms have been most helpful; William Goodman and Raphael Salaman, whose works on the history, nomenclature, and function of tools served as basic reference resources; the officers and trustees of the University of Vermont, who provided sabbatical leave for study and writing; many colleagues of the University, including Professor Branimir Von Turkovich, the staff of the Guy W. Bailey Library, and machinists of the Instrumentation and Model Facility; and friends in the Early American Industries Association who have contributed to my knowledge of tools.

Most of the tools that illustrate this book are from my own collection, that of Dudley Witney, and from other private collections whose owners generously loaned them for photography. Dudley Witney joins me in expressing gratitude for loans from the tool collections of J. Lee Murray, Jr., John S. Kebabian, Jonathan and Gail Holstein, and the late Milton J. Ernstof.

Permissions for photography were graciously granted by Stanley Tools Division, The Stanley Works (exploded view of "Bailey" iron plane); Colonial Williamsburg Foundation (coopering, and the cooper's, wheelwright's, and cabinet-maker's shops); Old Sturbridge Village (view of the cooper's shop); Special Collections Department, Guy W. Bailey Library, University of Vermont (illustrations from Nicholson File Company catalog, Joseph Moxon's *Mechanick Exercises*, and Asher Benjamin's *The Architect; or, Practical House Carpenter*); Early American Industries Association (plates from Joseph Smith's *Key to the Manufactories of Sheffield*); John S. Kebabian (Whipple, and Welch and Griffiths broadsides); and Mrs. Milton Ernstof (plate from Thomas Martin's *Circle of the Mechanical Arts*).

Paul B. Kebabian

AMERICAN
WOODWORKING TOOLS

FIGURE 1. Tools of the cooper, in shop at Colonial Williamsburg.

I
THE CULTURAL
HERITAGE

AMERICA WAS built by the hand tools of the woodworker. This statement is only a slight exaggeration: the tools that were an indispensable part of the meager baggage of the early settlers, the simple implements created by the local blacksmith or produced by the first small manufacturers, the later tools with ingenious improvements and refinements — all were essential to America's survival and growth. They were used to shape and build wooden objects and structures as diverse as cups and dinner plates, bedsteads and tables, buckets and barrels, carriages and farm wagons, houses and barns, bridges, mills, canal boats, and clipper ships. Only in the past fifty to seventy-five years has the hand tool in all its varieties of design and function been almost completely supplanted by the machine, becoming obsolescent and disappearing from everyday use in the woodworking trades. The tools that survive are witnesses to a time when craftsmanship was important, and when the satisfaction it brought to the artisan could be measured by the quality and individuality of his work. Today those tools our forefathers used to hew and saw, to shape and build, are archaeological artifacts — a significant part of the record of early American industrial life.

The function of a great many early American hand tools is clearly understood today. A limited number are still in general use, in forms that have existed for centuries. But the study of even these familiar tools is only in its infancy in terms of the identification of the individuals and manufacturing firms who produced them, the locales of their manufacture and use, the work accomplished by many specialized versions, and the place of these tools in economic history. The purpose of the study and research of woodworking tools is to attain and preserve knowledge for the light it throws on our past, as well as to conserve the implements themselves.

There are many limitations to such study. Although quantities of tools have survived, thousands upon thousands have been consigned — because of their obsolescence, their owner's indifference, or even patriotism — to the fire, the town dump, or to a wartime scrap metal drive. Some were simply used up; but in the hands of a careful workman who valued the tools of his trade, relatively few actually wore out or had to be discarded because of accidental breakage. The lumberman whose axe blade was reduced by wear and successive sharpenings over a period of years, or whose axe poll was damaged by being struck with a sledge, had the blacksmith re-steel the cutting edge or weld new steel to the poll. Such tools often show evidence of a succession of reconditioning jobs. For the most part tools were abandoned or destroyed because they became obsolete — as metals and other materials took the place of wood in their manufacture, and as the machine supplanted the hand of man in the woodworking processes.

In *Historical Archaeology*, Ivor Noël Hume states the case for the preservation and study of the artifacts from colonial and subsequent periods of American history. These objects are not antiquities in the sense that Egyptian or Roman artifacts are, but condescension is inappropriate; they have relative antiquity for us, and the future, one hopes, will be longer than the past. Yet the materials of our past, Noël Hume points out, are being lost and destroyed at a progressively more rapid rate. "The sound of the bulldozer is loud in the land."[1] For much of the historical record, it is a question of preservation now or never.

Destruction of the primary material is one major problem in the study of early American tools. A second is the lack of a large body of printed, manuscript, or pictorial records. Although it is clear that many sources are as yet untapped, the fact remains that much of the paper record has perished. It is gone because of our indifference and our inability to recognize its importance for the understanding of our past. In the hurly-burly of a rapidly developing nation, with the emphasis placed on innovation, expansion, and on the search for faster, cheaper, and better ways to create a product or do a job, the records of defunct manufacturing firms went the way of their obsolete products. Many early town records were victims of fire, flood, and theft. Great numbers of catalogs, advertising broadsheets, and other paper items of trade literature have disappeared. Contemporary manufacturing production records, accounts of purchasing of raw and finished materials, and data on sales, employment, and wages were generally considered of only transitory interest and usefulness, and were periodically discarded.

Growing maturity in understanding what may be historically useful has resulted in city and university libraries recognizing, as local

FIGURE 2. Homemade product of a sometime carpenter. French-Canadian match plane used to form a tongue on the edge of boards, such as Quebec vertical barn siding. Length, 13 inches.

FIGURE 3.   Double match plane. A French-Canadian carpenter fashioned a serviceable tool. One iron cuts the groove, the other, the matching tongue. Length, 9¾
inches.

and state historical societies did many years ago, the value of business and trade literature for industrial research. The development of archival collections of manuscripts, business records, ephemeral printed items and photographs, historical records surveys, and the registration of historic sites and buildings are all steps that contribute to the preservation of the larger historical record.

For the last century interest in the study and preservation of the materials of the American past has been increasing steadily. National, regional, and local associations whose aims are the preservation, study, and dissemination of information on early American tools have been established within recent years. Their membership has largely been drawn from interested and concerned amateurs. Although preservation in the sense of collecting has been a strong motivation among the members of these groups, there is a growing appreciation of the scholarly aspects. A desire to know who made a tool, where, and when, and how it was made, and for what purpose, is supplementing and displacing the mere satisfaction of possession. Such knowledge in no way conflicts with the collector's appreciation of a tool's intrinsic artistic and decorative qualities, its fine workmanship, the color, grain, and patina of its wood, nor with his admiration for the primitive, yet perfectly satisfactory, design of more modest examples of the tool-maker's craft and his ingenious make-dos.

Most early American woodworking tools adhere to tradition in form, design, and materials; the maker was not seeking, like an artist, to express an imaginative concept, to create an original form. Thus tools do not conform to some definitions of art. But whatever one's definition, one must concede that a tool made to meet a simple need can indeed possess aesthetic values. Through its shape, modeling, or line, through its decoration or the selection and use of materials, a hand tool can become an object of sculptural beauty. Observing and handling a specific implement, one's subjective, aesthetic judgment may quite often declare: "This is art."

Recognition of tools as artifacts of our his-

FIGURE 4.   Five bits for boring wood and metal. Forged by blacksmiths from worn-out files, typifying the forgotten concept of "use it up and make it do."

torical and cultural heritage has been one major factor in the recent growing concern for their preservation. They are displayed and used in living, interpretive demonstration by many museums and historical restorations. Until recently these institutions have used the actual old tools and implements in their demonstrations. But as curators have become increasingly aware that the originals were being consumed through use, reproductions made specifically for demonstration, and for use and handling by the visiting public, have been substituted.

Museums have collected tools for their aesthetic qualities, for their association with industrial history, and as examples of superior manual craftsmanship. Of course not all toolmakers displayed high levels of technical competence — mediocre workmanship is not confined to the twentieth century. One finds in artifacts made long ago both the skilled and the unskilled hand. But often the primitive tool, made by a workman who was unable or unwilling to purchase the manufactured product, reveals incidentally the artistic sense of its maker. It can be an object of great interest for the story it tells. With a basic understanding of the mechanical design and function of the tool, and within the limitations of his technical competence, his materials, and his lack of tool-making tools, the artisan was able to create a handmade version to fill his need (Figures 3, 5, 6).

While the purpose of most woodworking tools is known today, the use of some is still uncertain. There is, of course, a whole category

FIGURES 5, 6.   Construction lines. Plow plane shows scribed lines on both sides of stock. They guided the craftsman in using his chisels to shape openings for wedge, as arms of the fence, and depth stop.

of one-of-a-kind implements and fixtures whose function can often only be imperfectly deduced or sometimes defies explanation. These devices were made to do a particular job. Without knowledge of the activity or other implements or products the tool was associated with, many of these "whatsits" may never be precisely identified. And apart from such oddities there remain some tools once in common use whose function is now in question. It is only slightly more than a half century since the passing of the horse-drawn coach. The tools of the coach-maker are generally known and identified from printed sources and by provenance. Yet how and in what stage of building and finishing a carriage some of these tools were used is a matter of speculation and awaits further study.

It is encouraging to find that the collecting and study of both handcrafted and manufactured woodworking tools is becoming an important aspect of the new interest in the history of American technology. Investigation of these hand tools is hampered by the same difficulties that challenge the wider study of all of our material culture: the destruction of artifacts, the abandonment of industrial sites, the loss of ephemeral records, the absence of written records of techniques, and the disappearance of the oral traditions and know-how that once passed from master to apprentice. It is some compensation that these very difficulties alert us to the necessity of preserving what materials and records do remain.

The earliest tools used in America were brought from England, France, Holland, and Spain by the first settlers. Craftsmen carried their tools with them, and further importation began immediately thereafter. No census of seventeenth- and eighteenth-century tools surviving in America has yet been undertaken, but the majority of examples the author has seen that are not of American origin (and that bear identified tool-makers' names) are English.

A second early source for tools in America

was the blacksmith, who first fashioned them to order. At forge and anvil he could produce the axe, adz, drawing knife, or other iron implement needed by a neighbor. He also forged the cutting blade for the carpenter's plane or the cooper's howel. The blacksmith was the first tool-maker of modern times. Within his varied repertoire of products were edge tools, iron parts — such as rings for heavy mallets and wagon wheel hubs, hardware for wagons, spikes and ship fittings — and hinges, locks, and other household ironmongery.

For the most part, carpenters, sawyers, joiners, and other woodworkers of the seventeenth and earlier centuries made their own wooden stocks and handles to hold the iron shaping and cutting edges provided by the blacksmiths. There is no positive evidence to the contrary until about 1700, when plane-making became a specialized occupation in England. Certain members of the Joiners' Company of London began making planes not only for their own use, but also for the use of fellow guild members. They were the first tool-makers to identify their products by embossing their names on the wood with a metal die (Plate 4). The English tool-maker in iron also began to identify his work about this time, stamping his name or mark in the heated, softened metal of plane blades, files, and other edge tools (Figure 8).[2]

Several important conclusions can be drawn from this early product identification. It indicates a major change in the source for hand tools, and in the economic demand for them. Manufacture for sale and distribution was taking the place of a more casual, one-at-a-time production on individual order. The craftsman was no longer necessarily making his own tools, but rather buying them, trading for them, or in the case of the woodworking apprentice, often receiving them from the master as a condition of the successful completion of his apprenticeship. The maker's identification of his product was also a means of advertising. Shortly after names appear, the town place-name was often added.

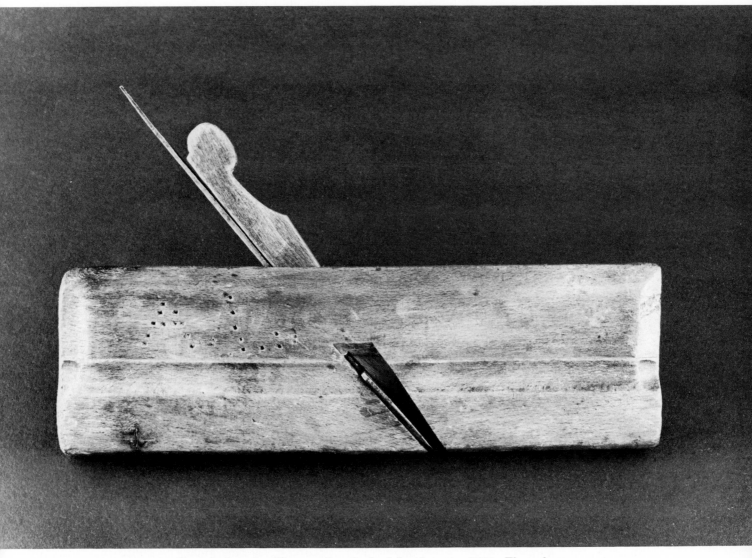

FIGURE 7. Round plane by Thomas Granford of London, ca. 1700. First plane-maker known to identify his product. Length, 10⁵⁄₁₆ inches.

FIGURE 8. Makers' marks. *Above:* Chisel by James Cam. *Below:* Plane iron by Weldon. Both were edge-tool makers of Sheffield, England.

This suggests distribution of the product beyond the tool-maker's immediate locale. In some cases the stamping of the name constituted a legal manufacturing or sales warrant, and we find acts of colonial and state legislatures in America requiring the artisan to identify his products by branding or stamping his name on them. Although tools with makers' names did not appear until about 1700, we know that tools in quantity — "bundles" of saws, for example — were shipped to the first American settlements. We must therefore conclude that tools were being systematically "manufactured" in England early in the seventeenth century.

The American tool-makers and woodworkers brought to their crafts a background of skills and knowledge akin to that of their English counterparts. The trades in seventeenth-century

America developed, however, within a substantially different economic system. From the later Middle Ages, the guild, or "company," was the formal craft organization of continental Europe and England. The guilds represented a movement toward a voluntary association of individuals, which ultimately brought together craftsmen of the same trade for both social activity and mutual economic benefits. The craft guild was a corporate structure, with elected officers and rules for the conduct of its business, which maintained strict regulation of the trade in its specialty. It fixed wages and prices for work, controlled membership by means of apprenticeship indentures, and sought monopoly control of production by the exclusion from the marketplace of articles not made by its own members.

10

The guild was the main channel for carrying forward a craft by providing for the training of an apprentice until he was sufficiently skilled to attain the status of journeyman, with the potential of becoming a master of his craft. Implicit in the word "craft" is the workingman's dexterity and skill, his subtle understanding of the feel of a tool in his hand, and knowledge of how that tool reacts to various densities, directions of grain, and other properties of the material being worked. Also implicit in the word are the private mysteries and technical methods of work which master passed on to apprentice — knowledge and skills which the members maintained in the strictest confidence within their own ranks.

Although the strength of the English companies was waning by the 1600s, they were still the dominant organization for control and management of the trades. Of the early settlers of America who were craftsmen, many had been apprenticed within the English system. The labor hierarchy of that system was carried over into the new colonial environment to some extent, with retention of categories such as apprentice, journeyman, and master, but the guild system in the English sense rarely became operative and never became institutionalized in America. The rapid growth of colonial towns, greater mobility of individuals, a high demand for skilled craftsmen, and a climate of enlarged personal freedom were not hospitable to the establishment of structured companies of artisans on the English model. Apprenticeship for the learning experience existed, but apprentices often served less than the usual seven-year term; the master to whom they were apprenticed was, more often than not, merely an independent freeman who practiced a trade. In the thirty-two years between 1695 and 1727, only one hundred and thirty-two indentures of trade apprenticeship were recorded in New York City, and by the mid-eighteenth century this form of labor contract (though still recognized by statute) was rapidly being abandoned.[3]

The tools used by artisans in the woodworking trades during the American colonial period can occasionally be identified in early manuscripts: correspondence, wills, estate inventories, ships' manifests, and customhouse records are all useful sources. Written or published information on how tools were used, however, is understandably almost nonexistent. The craftsman learned his trade as an apprentice or by experience on the job, with technical knowledge and the tricks of the trade passed on to him orally. There were no manuals published locally on how to use a particular tool, and were there any it would have been an unusual artisan who could have read them.

Although handbooks on woodworking and other trades did not become generally available until the nineteenth century, one notable exception was the publication in London, beginning in January 1677, of Joseph Moxon's *Mechanick Exercises, or The Doctrine of Handy-Works*. The book represents a significant contribution to present-day knowledge of seventeenth-century practices of the carpenter, joiner, blacksmith, wood turner, and other craftsmen, for Moxon describes the tools used in these occupations and explains how they were used; the specialized technical vocabulary of the trade is given in a glossary that completes the treatment of each subject.

Mechanick Exercises is the first basic "how-to" book. The preface opens with a statement of the author's recognition of the dignity of manual work, and his proposal to reveal the secrets of the trades. With knowledge of the rules and with "ingenuity and diligence" in their application, the author states that one could "inure his hand to the Cunning or Craft of working like a Handy-Craft." Moxon, to be sure, was not writing for the workingman but was addressing the gentleman-reader of his day, and he would hardly have expected this reader to learn the trade of carpenter or joiner simply from a published text. Nor, when he went to the various shops as an investigating reporter, did

IMPORTANT TO MILL-OWNERS.

WHIPPLE'S RECIPROCATING SAW-MILL.

Fig. 1

Fig. 2

PATENTED JANUARY 18, 1857.

This MILL possesses advantages far superior to any other now in use :

First, The Saw is perfectly strained, and the Saw-Frame is dispensed with.

Second, The Sweep or Pitman is dispensed with.

Third, The cost of construction is less.

Fourth, One-half of the power is saved.

Fifth, Such an advantageous motion is given to the saw that it will cut more lumber in a given time.

Sixth, It will require less expense to keep it in repair, as all the moving parts turn upon centers, thus avoiding the ordinary friction of slides.

Seventh, It is very compact, and can be constructed entirely portable, so as to be easily moved from one place to another.

Eighth, Only four feet space is required under the floor.

No better recommendation can be offered to those about to erect Mills, than to have them examine the principle for themselves. It can be applied to a JIG SAW, for shop use, or to a single Saw or gang for cutting timber and boards.

The Patentee is prepared to sell rights for territory, and to furnish all Iron Work necessary for the mill, with instructions to put it in operation, or furnish all running machinery for the mill, complete.

PRICE from $65 to $450, including the right of use in any one place.

CARLYLE WHIPPLE, Lewiston, Maine.

FIGURE 9.   Wood in the age of iron. Mill saw with mechanical power takeoff and iron parts, yet still framed in wood.

he get the whole story. But such criticisms do not alter the fact that Moxon did succeed in making public some of the "mysteries" of the trades, techniques carefully guarded by the masters and journeymen of the craft guilds.

Because the tools used in the early years of the settlement of America were the same as those Moxon describes, he is an invaluable source for our knowledge of seventeenth-century tools and techniques of woodworking. Two other major sources were published in the following century in Paris: the *Encyclopédie* of Denis Diderot, which describes and illustrates the sciences, arts, and trades (1751–1772); and André Jacob Roubo's *L'Art du Menuisier*, a masterful work on all branches of joinery (1760–1774).

The American hand tool for woodworking flowered in a country rich in wood, and in an age of wood — the period of less than four hundred years between the first permanent settlements to the close of the nineteenth century. We tend to think of the nineteenth century as an era in which iron was predominant, much as we associate plastics with manufactured products of today. In that century America did indeed become a major producer of iron and steel for use in machinery and in the transportation industries, and for a host of other applications as varied as stamped metal ceilings and decorative cast-iron fronts for buildings. But, important as metals were, wood was America's chief natural resource and a major component of virtually all things manufactured or constructed by hand well into the latter half of the century. In 1851 the jury examining the products displayed at the London Exhibition of the Works of Industry of All Nations — the "Crystal Palace Exhibition" — took special note of the American use of wood: "In America . . . machinery for working in wood is even more largely employed than with us. . . . The style of framing and designing these machines will at once betray their Transatlantic origin, and

exhibits great ingenuity, simplicity, and fitness for the purpose. They are principally framed of wood."[4]

The timber stands of the eastern seacoast of North America seemed limitless to the early settlers. Accounts of inland travel describe the dense and almost impenetrable forests that travelers and pioneering farmers encountered. Clearing land to grow crops was a first major undertaking, and one which incidentally provided a source of income. For after the timber was used to construct a small home, a shed, or a barn, the trees felled in clearing were usually burned, and the potash leached from the ashes became a cash crop or was traded for other necessities.

Wood was a prime export from the very early years of settlement. The General Court of the Company of Massachusetts Bay (which sat in London during the first years of the colony's establishment) in its session of September 29, 1629, assigned to Mr. Nathaniel Wright the responsibility for the sale of "clapboard and other wood" recently shipped from the Bay Colony to England. These clapboards had undoubtedly been riven from logs by hand with a splitting tool and mallet, although preparations were being made in that same year to set up the first sawmill in the Massachusetts plantations.[5]

Lumber exports increased rapidly. In 1770 over forty-two million board feet of lumber were exported from the colonies, the largest part going to the West Indies and the balance to England.[6] Not only large baulks, planks, and boards, but entire prefabricated house frames, the timbers knocked down for shipment, were transported from New England and Louisiana to the trading ports of the West Indies. Tench Coxe recorded almost two hundred "frames of houses" shipped in 1791/92, most of them made in Massachusetts.[7] Production for building the nation as well as for export increased greatly, and in 1799 four hundred million board feet of lumber were produced in the United

FIGURE 10. A basic concept. Smooth plane consisting of three parts: stock, wedge, and cutting iron. Canadian, made by hand with a few simple tools. Length 9½ inches.

States. By 1849, four billion feet of softwood and one and a third billion feet of hardwood were cut. Early in the twentieth century production reached a staggering total of forty-six billion board feet.[8]

In utilizing these vast quantities of wood to build America some hand tools retained the same materials of manufacture, the same basic construction, and the same applications over three centuries; others were considerably refined and specialized. The woodworking tool reached the culmination of its development in the second half of the nineteenth century, with a wide range of improvements. Changes in style, such as the provision of a rear handle grip on certain planes, made it easier for the craftsman to direct the tool. Modifications of mechanical operation, such as the shell chuck to hold the bit and the reversible ratchet, afforded a flexibility that the bit brace had lacked for centuries. Functional changes came about through the development of combination tools — one tool designed to per-

form the functions of two or more, often by the addition of fitted parts.

Nineteenth-century inventors applied their ingenuity to make tools work better, to allow easier and more precise adjustments, or at times to effect simplification in manufacture and a reduction in cost. On occasion there was an unfortunate tendency for human ingenuity to overreach itself, with inventiveness for its own sake producing excesses. Some combination tools were therefore poor substitutes for the individual implements they were intended to replace; nonetheless, a number were functionally quite practical. Often, of course, the simplicity of a prototype was sacrificed to achieve greater precision in mechanical movements or adjustments. The number of patents for improvements in hand tools mushroomed, although by the third quarter of the century it was relatively small compared to the total for woodworking machines. Between 1790 and 1873, over one hundred and twenty-five patents were granted

PRICES OF PLANE PARTS
"BAILEY" IRON PLANES

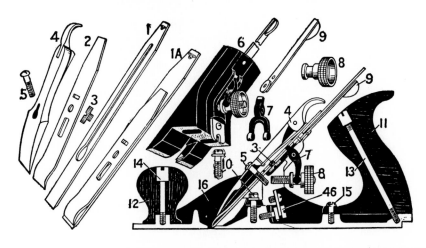

No	Name of Part	No. of Plane	1 2 2C	3 3C	A4 4 4C	4½ 4½C	A5 5 5C	5¼ 5¼C	5½ 5½C	A6 6 6C	7 7C	8 8C
1A	Double Plane Iron	....	.90	1.00	1.10	1.25	1.10	1.00	1.20	1.25	1.25	1.30
1	Single " "	....	.55	.60	.65	.80	.65	.60	.75	.80	.80	.80
2	Plane Iron Cap	....	.35	.40	.45	.45	.45	.40	.45	.45	.45	.50
3	Cap Screw	....	.10									
4	Lever Cap	....	.50									
5	" " Screw	....	.10									
6	Frog Complete	....	.70									
7	"Y" Adjusting Lever	....	.10									
8	Adjusting Nut	....	.20			for all numbers						
9	Lateral Adjusting Lever	......	.20									
10	Frog Screw	....	.10									
11	Plane Handle	....	.40									
12	" (Knob	....	.30									
13	Handle Bolt and Nut	....	.20									
14	Knob " " "	....	.20									
15	Plane Handle Screw	....	.....	.....	.....	.10	.10	.10	.10	.10	.10	.10
16	" Bottom	....	1.70	2.00	2.00	2.40	2.40	2.40	2.40	3.30	4.70	5.70
46	Frog Adjusting Screw	....	.10	.10	.10	.10	.10	.10	.10	.10	.10	.10

Add 10 per cent. for Corrugated Bottoms.
Add 30 per cent. for Bottoms and Frogs for Planes A4, A5, A6.

FIGURE 11. The concept developed. Parts for a smooth plane (from late 1800s to present) in cast iron, steel, brass, and wood. The "plane-maker" has become foundryman, machinist, and assembler of interchangeable parts.

FIGURE 12. Combination tool. The "jack-of-all-trades-and-master-of-none" tool. Made in Philadelphia in the 1880s, it provides a pliers, leather punch, wire cutter, box hammer, pipe wrench, hand vise, screwdriver, and tack puller. Length, 11 inches.

by the United States Patent Office for a tool as basic as the auger; patents were issued for the design of the bit, methods of attaching the handle, and for special technology in its manufacture. In the same period there were about two score patents on the hammer, thirty-five on the screwdriver, and two hundred on planes for carpenters, coopers, and joiners.[9]

The introduction of new materials and methods of manufacture also brought changes in tools. The woodworking tool had for centuries consisted of a metal working edge or surface which served to cut, abrade, or strike, combined with a wooden handle or stock. By the second half of the nineteenth century, iron, steel, and other metals were liberally substituted for the wood in many hand tools. Advances in metallurgy and in metal fabricating and processing now permitted machine forging, casting, milling, and stamping parts formerly made from wood. And these new precision-machined metal parts made feasible many functional and mechanical innovations that had not been possible in wood.

But the same thirst and genius for technological change that brought constant improvement to the tools also brought to an end the era of the hand woodworking tool. Craftsmanship in the trades succumbed to the machine and factory processes of manufacture. The woodworking tools of the age of wood, although now outmoded, retain their importance as vital clues to the American past, both deserving and in need of preservation and study.

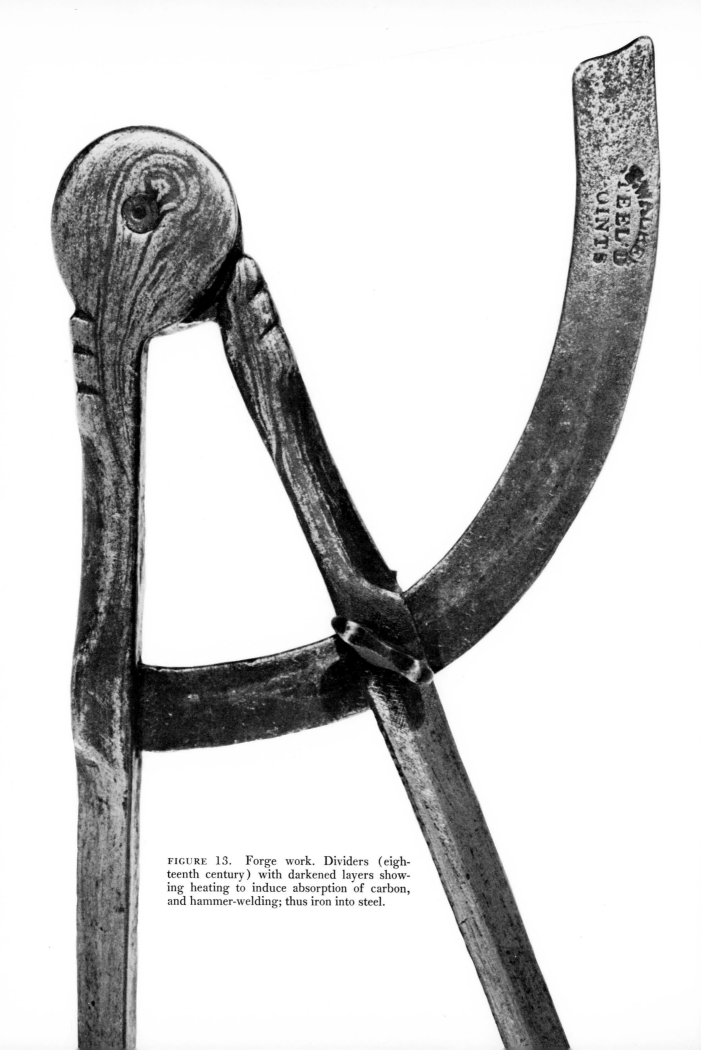

FIGURE 13. Forge work. Dividers (eighteenth century) with darkened layers showing heating to induce absorption of carbon, and hammer-welding; thus iron into steel.

II

HISTORICAL OVERVIEW

DISCOVERIES OF anthropologists in the past twenty years have moved back the appearance of man the tool-maker to between two and three million years ago. Among the most significant discoveries have been those made by Louis and Mary D. Leakey at Olduvai Gorge, in northern Tanzania. In that East African site between 1959 and 1972 they found cranial and other bones of the ancestors of modern man, and in 1974 Mary Leakey recovered further fossilized hominid bones which date from three million years ago. Among the implements located with these earliest known remains were hammer-stones, hand axes and chopping tools, chisel-edged implements, scrapers, cleavers, and tools similar to burins and awls for making grooves

and holes. These tools were made of lava, quartz, quartzite, and fragments of bone. Symmetrical, round stone balls also discovered at Olduvai Gorge were possibly bola stones, stones attached to cords or thongs made from animal skins, which were hurled at animals to entangle them. The implements of this age were chiefly associated with the acquisition of food and its removal from animal carcasses; initially these tools were either weapons used in hunting or tearing instruments which served the functions of teeth and claws of animals in stripping carcasses. Their makers displayed a conscious selectivity in use of stones of varying properties and color to make different types of implements.

THE AGES OF STONE AND METALS

Man's tool-making has been associated with periods identified by the materials from which his tools were fashioned: Stone Age, Bronze Age, and Iron Age. The Stone Age was that

unimaginably long period from the time man first made stone implements to the coming of the Bronze Age. The Bronze Age, evolving about 3500 B.C., was a period of rapid develop-

FIGURE 14. Stone Age hand adz. Alaskan Eskimo, sharp and highly polished, from dense jade. Length, 5 inches.

ment accompanying the beginnings of smelting and the manufacture of bronze objects. The Iron Age began with the discovery of iron and the technology to reduce ores to a product that could be forged into tools. Our present age of steel and the machine can be considered a part of the Iron Age.

These "ages" of man are in fact not strictly datable; they are relative cultural time frames. Thus, a Stone Age tool may be one made by the earliest tool-making man, by a nineteenth-century Australian aborigine, or by a twentieth-century Indian of the upper Amazon; the implements of each of these tool-makers are the product of a Stone Age culture.

Stone Age man made his core, flake, and blade tools by a variety of methods, including percussion (striking a stone core with a heavy, hand-held hammerstone or with a lighter bone or wooden "baton"); pressure flaking (splitting away flakes of stone with a pointed bone or stick); and grinding and polishing. In the Neolithic, or most recent, period of the Stone Age, stone tools were used for felling trees, for building, and for making canoes, sleds, spears, and other wooden implements. Thousands of stone tools have survived. It is difficult, however, to determine how extensively they were used to fashion objects of wood because the wood, exposed to air, to weathering, and to dampness, ultimately disintegrated. We may deduce that wood was in daily use from implements, such as the bow and spear of Stone Age man, unearthed from European peat bogs. The British anthropologist Kenneth Oakley suggests that many stone tools with hollowed, concave scraping edges would have been appropriate for shaping the rounded surfaces of these wooden implements.[1] Experimentation has taken place in the modern reproduction of Stone Age implements, and anthropologists have successfully duplicated both the processes of manufacture of the tools and their application in cutting, scraping, incising, and boring. These studies have been significant in testing theories of how the tools were made and how they functioned.

Much more exacting research into the making and use of tools by the Russian ethnographer S. A. Semenov has been based on the fact that many stone and bone tools carry not only the marks of how they were produced, but also how and on what materials they were used.[2] Microscopic examination and microphotography reveal the direction in which the tool worked and thus provide information on how it was held in the hand. Semenov has concluded that prehistoric man used wood extensively in the late Stone Age, when planting and harvesting of crops, fishing, and the domestication of animals had begun. As man became a food producer rather than a food gatherer, he gave up his nomadic life and established settlements. These produced a need for dwellings and for the tools to build them. Semenov's microanalysis shows that axes were used to fell trees, and that stone hand adzes were used in hollowing operations, probably for such work as shaping dugout canoes. The relatively rough cutting edges of the earlier Stone Age stone axe, adz, knife, and chisel, which for ages had been adequate for man's most primitive needs, were refined in order to cut more efficiently. Stone surfaces were ground and edges polished on sandstone blocks to create a surface and edge that reduced the resistance of the tool to the wood or other material being worked. Semenov supported his scientific conclusions on the use of the axe in felling trees with practical evidence: a fir tree ten inches in diameter was cut through in twenty minutes with a nephrite axe.

In 1960 the ethnographer Vladimír Kozák had the rare opportunity to observe the manufacture of a stone axe by a people living in a Stone Age culture.[3] He was a member of a scientific expedition in the Amazon region of Brazil which sought out a small, remote tribe of Indians known as the Héta. In fact the Héta had made their presence known when they saw a native Brazilian farmer clearing land in the wilderness, using a steel axe. They marveled at the efficiency of the modern tool, recognizing its superiority over their stone implement. Kozák

FIGURE 15. Bronze Age axes: *Above:* Late form, cast in a mold, with hollow socket at top to receive crooked wooden handle. Length, 5⅛ inches. *Below:* Earlier form, hammer hardened. Length, 5 inches.

prevailed on a member of the tribe to make a stone axe and recorded the entire process of manufacture. The tool-maker first carefully selected a suitable elongated ovoid stone; for several days he shaped it by removing fine particles with a hand-held hammerstone until he had roughly obtained the shape required for the blade. Then, in a combined grinding and polishing process, using a sandstone block, white clay, and water, he rubbed the blade until its smoothness and sharpness met his satisfaction. After selecting a four-foot length of hardwood about five inches in diameter for a handle, the Indian made a chisel from the leg bone of a tapir, which he also sharpened on the sandstone block. He next chiseled a deep oval hole in one end of the axe handle to receive the upper, unpolished end of the axe head. He trimmed the remainder of the handle to a diameter of about two inches by splitting off excess wood with the bone chisel, and finally force-fitted the axe head into the hole of the handle. With these stone axes the Héta felled trees as large as four feet in diameter.

Significant developments in the late Stone Age, all taking place very gradually, over thousands of years, materially improved man's tool-making processes and the effectiveness of the tools. The angle of a cutting edge was reduced by grinding to effect easier penetration. Friction created by the tool in cleaving or cutting was reduced by smoothing. A great leap forward was made in increasing the mechanical power of a tool by providing it with a handle; the effective radius of the swing was thereby increased to transmit more force to the working head. The bow drill was developed — possibly inspired by the mechanical action of a piece of wood twirled between the palms of the hands to start a fire by friction — providing a rapid rotation and counterrotation of the bit for more friction and thus more efficient drilling.

Man's progress as a maker of objects to fill his needs is continuously reflected in improvements in his tools. These changes are associated with a number of factors: the material from which the implement and its cutting edges were made; the hardness and strength of the material; the ease and speed with which the tool could be fabricated; the functional design; and the facility with which a worn instrument could be sharpened for re-use. Prehistoric man took relatively little time to fashion a rough core tool, such as a heavy chopping implement, but making a more refined flint or obsidian blade tool, by using a hammerstone or by pressure flaking, was more time-consuming. The sites where tools were made, such as the Stone Age "factory" discovered in northern Wales where many axes were fashioned from a particular greenish lava, provide a record of incomplete manufacture and the discarding of quantities of imperfect tools.

The evolution from the Stone Age to the ages of metals was of great significance in the history of man. The use of copper, bronze, and tin for tools, weapons, decorative ornaments, and household objects ushered in the Bronze Age. Sharper and more durable tools could be produced more quickly by technological advances such as work hardening, smelting and casting, forging, and heat treatment. Copper, the essential element, was first worked cold, by hammering, to make small ornamental objects such as the beads and pins found in an Egyptian grave from 4000 B.C.[4] Fully fired pottery dates from this same era, and the relationship between pottery kilns and confined fires for annealing and smelting metals is significant. The metallurgical technique of annealing, wherein large pieces of free copper were heated and cooled as they were worked and reworked by hammering to desired shapes, was an important discovery. Annealing prevents cracking and fragmentation of the copper, and hammering doubles the hardness of the native metal in the finished product. The earliest known copper tools, which include chisels, engraving tools, and saws, were found in a First Dynasty Egyptian tomb; they are five thousand years old.[5] Archaeologists assume that the use of copper originated in western Asia, in the regions south of the Caspian Sea and in upper Persia. Migration carried metal-working technology into the fertile crescent area of Mesopotamia, the lands at the eastern end of the Mediterranean Sea, and to Greece, Crete, and Egypt. Little copper was available in pure form, so its use could be developed only through the introduction of smelting. In this process, copper ores, in the form of malachite and azurite, blue ore-bearing stones, were reduced in the charcoal fire of a furnace and the refined copper separated from slag, ashes, and other materials of the ore.

Smelting originated about 4000 B.C. and within five hundred years the method had become widely known in the ancient world. Much of our knowledge of the early use of copper comes from artifacts made in the Tigris and Euphrates valleys of Mesopotamia by the Sumerians, who developed and used the smelting process extensively between 3000 and 2500 B.C. Because there were no copper ores in the valleys they farmed, they imported ores from the mountains to the north. The process of reducing copper ores to a relatively pure form involved resmelting to remove slag and other impurities, which were skimmed off the fluid metal. Once this process was developed, the Sumerians recognized the potential for casting the liquid metal into molds, to make tools, weapons, and ornaments. Closed molds of baked clay and stone were initially used for casting. The earlier free copper tools, made by hammering and annealing, tended to mimic their stone antecedents, but with casting, tools were created in new shapes and dimensions.

We may assume that the first bronze was produced by accident, probably not long after the smelting process came into general use. Bronze is a combination of copper and tin, in an approximate proportion of ten to one. The alloy becomes fluid at a lower temperature than does copper alone, making it easier to cast into tools, and it is substantially harder than copper. Tools with bronze cutting and shaping edges were more efficient than copper-edged tools. They could cut, chip, scrape, and bore faster, needed less frequent sharpening, and lasted longer. Bronze supplanted copper slowly, over several hundred years, during which both metals were in general use. Bronze Age cultures of varying durations encompassed virtually every part of the civilized world. Examples of bronze tools are found from Britain eastward to Russia, Siberia, and the Orient, where bronze casting became a highly developed art.

The intentional, rather than accidental, separate smelting of copper ores and tin ores to reduce them and subsequently combine them to make bronze probably took place between 2500 and 1500 B.C. The proportion of tin in implements before that time suggests that much "bronze" was reduced copper ore that happened to contain some tin. This could well apply to bronze axe heads and chisels from Ur in Mesopotamia (ca. 2600 B.C.) which reveal highly developed skill in tool making but considerable variation in the ratio of copper to tin.

Iron came into general use only three thousand years ago. The Hittite peoples, inhabiting a central area of Asia Minor, were the first to produce iron in quantity, by applying the smelting process to hematite, the source of iron ore in the mountains of that region. The Hittites were able to produce the conditions necessary to reduce iron ore: a charcoal fire in a stone furnace lined with clay, with a forced draft generating sufficient heat to allow the iron to be released from the ore. The process produced a "bloom," a spongy mass of iron particles mixed with slag and ashes, at the base of the furnace. The spongy aggregate was then reheated to bright red and hammered on an anvil, the process welding the iron particles together and forcing impurities out of the bloom. By repeated heating and hammering, the wrought iron was fashioned into bars suitable for working into tools and other articles. The reducing of iron ore requires a temperature considerably higher than that for smelting copper or tin. But once this process was mastered, iron — whose ores were widely distributed over the surface of the earth

— gained ascendancy over copper and bronze for tool-making.

Iron was a highly valued precious metal in the first centuries of its manufacture, worth five times as much as gold and forty times as much as silver. (An indication of its value was the presence of several iron tools in miniature in the tomb of the Egyptian pharaoh Tutankhamen, about 1350 B.C.) By about 1000 B.C., extensive mining and smelting operations existed in the areas south of the Caspian and Black Seas and in parts of the Middle East. Wrought iron was transported from these regions over the trade routes to the Aegean islands and lands bordering the Mediterranean. Between 750 B.C. and 50 B.C., Hallstatt in the Austrian Tyrol, and La Tène on the Lake of Neuchâtel, both early centers of Celtic culture, passed from the Bronze Age into the Iron Age. The Celtic conquest of Europe, from Spain to the Black Sea, resulted in the spread of their culture and in the wide diffusion of iron-making technology during the pre-Christian era.

To the south, the ancient city-state of Meroë, at the border of southern Egypt and the northern Sudan, flourishing from the sixth century B.C. to the fourth century A.D., became an iron-producing center. Iron formations in the neighboring sandstone hills provided an accessible source for ores, and the presence of six large mounds of slag at points in the present ruins of the town show where furnaces were in operation. Meroë was a focal point from which knowledge of metalworking techniques was carried westward on the African continent.[6]

Iron tools supplanted bronze almost completely after the discovery of two processes, carburizing and hardening. In ancient furnaces, iron took on some properties of steel by absorbing carbon from the charcoal fire in which it was directly reduced from ore. The blacksmith used the wrought iron, which was the end product of smelting, to shape tools by forging. Each time he reheated iron in his charcoal fire carburizing took place as more carbon penetrated into the outer surface of the iron, forming

a still stronger metal, steel. Awareness of the results of this process (rather than an understanding of the chemical change in the properties of the metal) led to intentional carburizing; the blacksmith then carburized thin strips of iron and welded them together in a bar of steel as stock for tool-making.

The first hardening process, known to have been in use about 800 B.C., was quenching — rapid cooling of iron by immersion in water. A most significant breakthrough, attributed to the Romans, was the development of tempering, an improvement on simple quench hardening. Steel could not be tempered until it had been hardened by heating it to a bright cherry-red color and then rapidly cooling it in a liquid. The blacksmith forged a tool from the steel and then tempered it by reheating it to the correct temperature to bring it to that degree of hardness and toughness required for the specific tool; files, chisels, axes, saw blades, and bits for drilling required different heats. After reheating, the tool was quenched by plunging it in water or other liquid.

When the ability to measure temperature accurately and knowledge of the properties of metal were limited, obtaining the desired result of the tempering process was largely accidental. But the ironworker discovered by experience that heating the metal to certain temperatures that he could identify by color — light straw, dark straw, light purple, or blue — produced the qualities required for a specific tool. Insufficient hardness would produce a tool that would deform or rapidly dull in use; excessive hardness would produce one that was too brittle and would chip and fracture.

From the Roman era to the late Middle Ages there were no basic changes in the manufacture of iron. Shape and construction of the furnace were altered to increase efficiency; hotter fires were obtained by using a "trompe," or air aspirator, to produce continuous air flow in place of the intermittent air supply of hand- or foot-operated bellows; mechanical trip hammers took the place of hand power in refining the bloom; but into the 1300s the product was still wrought iron. By the early fifteenth century a radical change had taken place. Furnaces over thirty feet high were built, and water-powered bellows, producing greatly increased air pressure, raised furnace temperatures to well over one thousand degrees centigrade. The smelting process for the first time produced liquid iron, and it was possible to run the metal from the forge into trenches on the ground, forming

"pigs" to be reworked in manufacturing, or to cast objects in molds.

The blacksmith in colonial America, like earlier ironworkers, developed his skill largely by trial and error. He gained a vast store of knowledge by his own experience and by learning techniques passed to him by older furnace and forge workers, but his technology was completely lacking in any theoretical underpinning.

EARLY DEVELOPMENT OF TOOLS

The determination of the fundamental functions served by the hand tool, as well as many of the basic designs, developed in prehistoric times. Certain physical motions of the hand, arm, and body are required to impart physical force to striking, smoothing, abrading, and other actions of a tool on wood. After millennia of working with stone tools, the motion patterns and angles of application of cutting edge to the wood were quite clearly understood by the Bronze and early Iron Ages. The tool designs which evolved incorporate those understandings. Once the basic form of the tool was determined, changes were made only in its materials, to improve and refine the tool action, or to provide a new method of holding the working edge to facilitate its use.

Early tools of copper, bronze, and iron have been recovered from widely scattered locations in the lands of ancient civilizations in the Mediterranean area, in Europe, and in Britain. The Egyptologist Sir William Flinders Petrie has compared Egyptian tools with similar ones from other regions, cultures, and eras.[7] Although the Egyptians had to import copper and tin, the range of their bronze woodworking tools

was extensive. Their earliest axes had heavy, plain blades with one or more holes at the top for lashings; the blade was inserted in a slot in the handle and lashed to it with leather thongs (Figures 16, 17). Later, the top of the blade was widened and lugs provided to bind it more firmly to the handle. Copper and bronze examples from Mesopotamia, Syria, and countries of southern Europe show that the socketed axe — the most satisfactory type for affixing the handle — was known before 2000 B.C., but Egyptians apparently never used this form.

In Crete, Greece, and Rome, the double-bitted bronze axe was developed — a tool with two symmetrical cutting edges and a handle socket formed at midpoint of the head. In Europe and Britain, however, makers of the bronze axe generally ignored the efficient socket, or "eye," positioned at right angles to the head, and made axes which required either a split wooden handle to which they lashed the head, or a crooked handle with the short end fitting into a hole at the top of the casting (Figure 15). Medieval iron axes — long, narrow-bladed felling axes, and various short-bladed forms — were forged to provide a more or less round

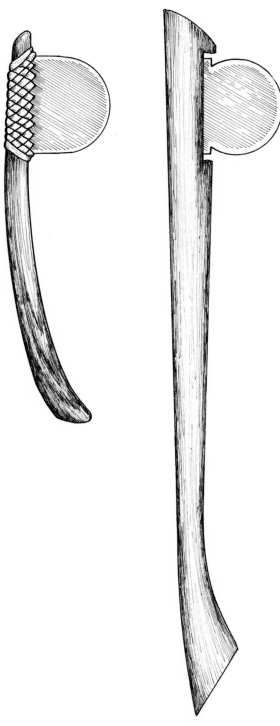

FIGURE 16. Egyptian axe, Third Dynasty (2700 B.C.). Bronze blade inserted in wooden handle and lashed. (After Petrie)

FIGURE 17. Egyptian axe, Twelfth Dynasty (1900 B.C.). Inserted bronze blade with lugs (lashing not shown). (After Petrie)

handle-socket positioned as in the tool made today.

The adz, which was hand held through eons in the Stone Age, became a hafted tool in the Bronze Age. Its blade was always at right angles to the handle, and as the tool ultimately evolved the blade was sharpened only on the inner edge (Figure 26). It performed a surface smoothing function, not unlike the action of the later plane, by removing thin shavings or chips. The blade was at first lashed to the handle with thongs, later held by a metal strap and wedge, and ultimately socketed.

The wooden club was a common tool form, predating metal percussion tools. Wooden striking tools were in use in the Bronze Age, for examples of such implements some 4500 years old have been preserved from the Fifth and Sixth Dynasties of Egypt. They were undoubtedly used in woodworking for striking chisels and gouges, and for similar work with stonecutting tools. The hammer form of tool in bronze was succeeded by the iron claw hammer with a handle, used first in Greece and Rome.

Almost all iron chisels and gouges made prior to the seventeenth century have rusted away or were remanufactured into another product when worn out, but earlier Bronze Age implements of this kind have survived from a number of locations in Egypt, Greece, Italy, and Britain. The early chisel was made entirely of hardened copper or bronze for working either wood or stone. The body was curved and the cutting edge of the blade was beveled equally on both sides. But to cut deep mortises with straight sides and clear the chips, the chisel had to be not only strong, but beveled on only one side of the cutting edge. This true chisel form was in use in Egypt by 2500 B.C.[8]

Stone Age shells and beads with holes drilled through them for stringing provide evidence of the antiquity of boring tools. The drilling implements used were fashioned from bone, reindeer antlers, and flint or other stone. With the introduction of metals, copper and bronze took the place of earlier materials, but well into the Bronze Age the Egyptians were performing precision boring with core bits fashioned from stone.

Implements for boring wood came of age when metals could be utilized for cutting bits. The early bow drill had a wooden shaft into which a metal cutting bit was inserted. By alternately pushing and pulling a wooden bow, which was strung with a thong or rope wound once around the drill shaft, the operator set up a reciprocal rotating motion of the shaft and its cutting bit. The bow drill appears frequently in Egyptian wall paintings depicting woodworkers (several examples from as early as 2000 B.C. are described and illustrated by Flinders Petrie). Using the drill, the worker held a rounded cap in the palm of his hand as a pivot bearing for the upper end of the drill shaft. These drill caps were made from stone, or from the nut of the Theban palm tree, which becomes exceedingly hard when dried. In a later Stone Age culture, the North American Eskimo fashioned his bow drill and bow from bone. A third piece of bone, with a small depression to function as a pivot bearing, he gripped with his teeth. This left one hand free to operate the bow and the other free to position the drill and hold the workpiece. Bow drills of this construction are still in use among the Eskimos of the Diomede Islands in the Bering Strait.

Small flints with chipped cutting edges were used in the Stone Age as saws, primarily to shape small objects from stone, using sand and water to assist the abrasive action. Some flint saws were flaked to a serrated edge, while others were made by inserting several chips of flint into a slot in a wooden stick, adhering them with asphalt. The flint chips in these saws were wedge shaped — wide at the top and tapered to a sharp edge — and so could penetrate to only a small depth.

The earliest flat woodworking saws were of hammer-hardened copper or bronze, with V-shaped teeth. These were abrading tools, operating on the pull stroke, with which the workman

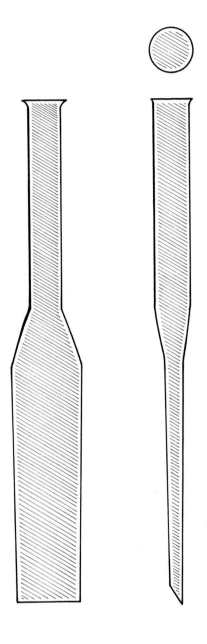

FIGURE 18. Bronze chisel. True form of chisel with round handle and flat blade sharpened on one side; an ancient Egyptian development. (After Petrie)

27

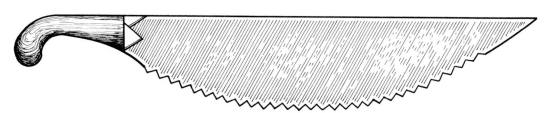

FIGURE 19. Egyptian saw, Sixth Dynasty (2330 B.C.). Copper or bronze, with V-shaped teeth, from a scene depicting sawing. (After Petrie)

laboriously removed particles of wood by pressure and friction. The Romans were the first to regularly rake saw teeth — filing them at an angle to effect a chipping action by each tooth — and also the first to set saw teeth. Pliny, writing in the first century A.D., refers to Roman saws with points of the teeth set, or bent, alternately right and left. This refinement results in a kerf, or saw cut, wider than the thickness of the blade and prevents the saw from binding as it cuts, particularly in unseasoned wood. The metal in these earliest saws was too soft to produce a thin, stiff, yet flexible blade, but by forging a relatively thick blade, the Romans were able to develop a saw that operated on the push stroke. (Saws of this type were made before A.D. 79, as examples were found in the ruins of Pompeii.) More strength can be applied in pushing the saw than in pulling, so the push saw cuts faster, but it is easier to maintain a straight and true cut with a pull stroke, and there is no opportunity for the saw blade to buckle.

Virtually all western hand saws since the Roman era have operated on the push stroke, and the points of the teeth are therefore raked toward the front end of the tool. The thwart saw, cutting on pull and push stroke by one man working at each end, is an exception; its teeth have been formed in a bewildering variety of patterns, with chipping action and clearing of sawdust usually effected by irregularities in the depth and set of the teeth. In some styles, the teeth of this lumberman's saw may be raked in two directions to effect the reciprocating cutting action. Japanese and other Oriental hand saws

of today still operate in the early Bronze Age way, on a pull stroke.

The plane is a tool of relatively recent origin, with no counterpart in either prehistoric or Bronze Age cultures. Over sixty years ago Petrie attributed the invention of the plane to the Romans, an attribution based on a modest number of surviving Roman examples and a quantity of separate plane cutting irons. Since Petrie's time, no artifacts have been discovered that would place the invention other than in the period of the Roman Empire. If the tool did originate in Greece, or elsewhere, the Romans seem to have been chiefly responsible for putting it to use. It has been suggested by Josef Greber that the plane developed from the hand adz: a wooden or metal stock with a flat sole could have been substituted for the handle, and an adz-form of blade strapped to the fore end.[9] There is, however, no clear evidence of such a transitional development.

Once again the destruction of Pompeii in A.D. 79 provided evidence for dating. Petrie recorded four planes found in the ruins and described their construction: "A continuous plate of iron goes from the back along the base, over the front and top, and is rivetted to the base plate at the back. The top opening is cut through this plate. A cross bolt, run through the wooden body, serves to block the wedge which holds the cutter in place."[10] In 1964, William Goodman provided a descriptive inventory of thirteen known Roman planes, though listing but two from Pompeii.[11] The stocks of eleven were made from iron and wood, with working surfaces made of iron. The planes listed vary in length from six and one-quarter inches to seventeen and three-eighths inches. They were made with integral handles cut in the stock for gripping and directing the tool; in the shorter lengths one handle was usually provided in back of the cutting iron; in the longer, two handles, one in front and one to the rear of the iron. Although only two were made entirely of wood, the number of surviving Roman plane irons — Greber listed fifty-nine in 1956[12] — suggests that it is very likely that the usual plane was made with a wooden stock to hold the cutting iron, and that this type has all but completely disappeared because the wood decomposed.

The medieval period and succeeding centuries to the 1600s are truly "dark ages" as far as our understanding of the uses then made of many woodworking tools. Certainly the crafts persisted throughout the periods of plague, war, famine, and political and social anarchy which characterized much of the Middle Ages. Tools were needed to build fortified castles, churches, monasteries, and houses; to build ships for trade and for naval warfare; and to beat out arms and armor at the forge to supply local and holy wars. But relatively few examples of those tools have survived, and often the best evidence comes from pictorial representations; this is particularly true for planes. Medieval axes, saws, augers, and a variety of shipbuilding tools in metal have been recovered. In almost all instances the wood portions, such as the handles, have disappeared.

The guilds, which from the twelfth century became the organized associations of craftsmen, fostered a tradition-bound system and did little to engender innovation either in the development of tools or in new methods of performing work. New types of tools were, of course, developed in the centuries following the Renaissance, but radical changes in hand tools were not to take place until the nineteenth century, when new materials, altered tool-manufacturing processes, and inventive mechanical changes were introduced.

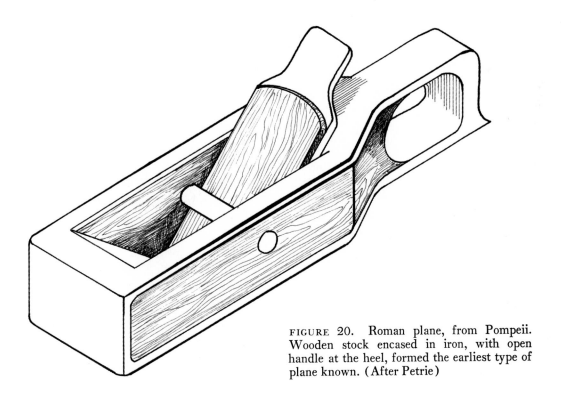

FIGURE 20. Roman plane, from Pompeii. Wooden stock encased in iron, with open handle at the heel, formed the earliest type of plane known. (After Petrie)

·-WOODWORKING TECHNIQUES-·

From ancient Egyptian wall paintings and objects found in tombs it is apparent that woodworking occupations — the crafts of wheelwright, chariot-maker, coffin-maker, carpenter, and cabinetmaker — were of great importance to the Bronze Age Egyptians. Although the supply of native wood was limited, many objects of native sycamore, palm, tamarisk, and acacia have survived. Acacia was the only local wood suitable for furniture and cabinetmaking. Ebony was imported, so highly valued that it was received by the pharaohs as tribute; it was used in royal furniture and cabinetwork. The finest examples of woodwork, often inlaid with ivory or decorated with gold, were traditionally placed in the tombs of rulers and high state officials.

Techniques of joinery were well developed, and a number of methods were used. Although boards were sometimes joined by the simple expedient of drilling holes along the edges and lashing with leather or fiber thongs, more refined techniques were also mastered. Flat tongues, or keys, rather than round dowels, were sometimes used in joining boards. These

keys were about two inches wide and were inserted in surface mortises chiseled into the abutting edges. To secure the keys, a hole was drilled through both board and key on either side of the joint, and a small round wooden peg driven through.

Furniture from the tomb of Tutankhamen is remarkable for the evidence it provides of skill in woodworking techniques. The frames of both a ceremonial chair and a child's chair are made with mortise-and-tenon joints, through-pegged with bronze or copper rivets capped with gold, and the five curved wooden slats of the seats are joined in the same way. A small gaming table of the pharaoh has a long drawer, the drawer front joined to the sides by dovetail joints that are identical to cabinetmaker's dovetails of today. These examples also reveal the art of skillful wood-carvers; the legs are shaped in naturalistic representation of the legs of a lion, with carved claw feet — some with inlaid claws of carved ivory. A portable chest is made with a frame of ebony, and recessed panels (possibly of cedar) form the side panels; a veneer border of four strips, alternating ebony and ivory, surround each panel, and the lid rests on top edges of the chest which are shaped to a simple wide cove molding. The dovetail joint, mortise-and-tenon joint strengthened by dowel, framed panels, decorative carving, molding, veneering, and inlay work — a catalog of woodworking techniques required for the finest cabinetmaking of the eighteenth and nineteenth centuries — were being used at least three thousand years ago.

Although we do not know what tool was used to form a molding at the time of Tutankhamen, within five hundred years — between 900 B.C. and 700 B.C. — many implements had taken the forms that were to persist to the zenith of the hand tool in the 1800s. "The most perfect forms of bronze chisels were then devised," Petrie writes, with both tang and socket handles in wide paring forms and deep mortise shapes; and hand saws, rasps, files, and wood bits with a projecting center point to effect straight-line boring were all perfected.[13] Not only was the basic design of many metal-edged tools thus established, but a variety sufficient for the fundamental techniques of woodworking and joinery was available in ancient lands when civilization was only first reaching the continent of Europe.

FIGURE 21. Boy's tool chest. The illustration on the cover shows boys constructing a rowboat — but not with these tools. Many youthful would-be woodworkers were undoubtedly frustrated by such ineffective toy tools as these, purchased by well-intentioned parents.

III
SOME BASIC TOOLS

THE CLASSIFICATION and description of tools presents some difficulties. One may group them by their function as percussive, abrading, drilling, or smoothing implements, or classify them on the basis of the trade with which they are associated. We have chosen to discuss in this chapter a number of tools that have been used in a variety of American trades, and to treat in later chapters tools associated primarily with one specific occupation. This method is at best a compromise. The saw, for example, is used in many branches of woodworking; in some form it is at home equally among the tools of the carpenter, the wheelwright, the cabinetmaker, the shipwright, the coach- or wagon-maker, or among those of a specialized craftsman such as the stair-builder. The adz is also used by a variety of workers in wood; specialized forms would have been used by the carpenter, the shipwright, the cooper, or the farmer making a wooden kitchen bowl for his wife.

During the entire period of American history under consideration, many households, particularly the country home and farm, would have kept a variety of general-purpose woodworking tools for routine carpentry and maintenance, as well as some specialized implements for seasonal occupations. Although the rural and small-town economy of the eighteenth century supported a number of specialized artisans, the majority of the population farmed. And each farmer had to be something of a jack-of-all-trades: he was wagon-maker; house- and barn-builder; maker of hats, cloth, tools, furniture, nails and handspikes, staves and heading and hoops for barrels, potash, maple sugar, or any of a host of other products for use at home and for trade or sale. Many farmers also had occupational sidelines. Writing about communities he passed through on a trip from Boston to New York in 1788, a French visitor, Brissot de Warville, remarked that "almost all these houses are inhabited by men who are both cultivators and artizans; one is a tanner, another a shoemaker, another sells goods; but all are farmers."[1]

We will here consider the function and manufacture of a group of basic woodworking tools common to a variety of occupations: the axe, adz, hammer, chisel, saw, boring implements, plane, and drawing knife.

THE AXE

The English and American felling axe of the seventeenth century was fashioned of iron, and weighed from three to five pounds. Like the majority of woodworking tools, the axe is a composite implement, consisting of head and handle. The broad sides of the head taper equally on both sides to a knife edge. The felling axe is used with a two-handed overhead, over-the-shoulder, or side-swinging action to deliver chopping strokes, primarily across the grain of wood, such as that of a tree trunk. An early axe of different design, the broad axe, was normally fashioned with one side tapering to a long basiled, or beveled, cutting edge and the other side made relatively flat. This axe was used with short, two-handed strokes as a shaping tool to hew wood, working with the grain, as in smoothing the sides of a log to form a square timber. A short handle, bent laterally (outward from the beveled side of the axe head) allowed the broad axe to be used with a paring action, close to the work.

The axe head has been forged in a wide variety of forms: roughly rectangular; flared from the top to the cutting edge; flared in a "goosewing" pattern; narrow bladed to shape a mortise; and even with a wide, incurving cutting edge for axes used to slash pine trees for extraction of turpentine oleoresins. The eye in which the handle, or helve, is secured was round in early versions but subsequently was shaped in oval or teardrop form.

The first felling axes used in the settlements of America were English or European importations, but axe-making shortly became a regular occupation of the local blacksmith or plantation mechanic. According to Victor Clark, in his *History of Manufactures in the United States*, the axe was probably the first tool made in America for general sale. This was made possible by an infant iron industry, which was pro-

moted and encouraged in the colonies from an early date. In 1644 the Massachusetts Bay Company granted a twenty-one-year exclusive right to establish an ironworks, with a ten-year tax-free status. The "undertakers" of the venture (who included John Winthrop, Jr.), had to buy shares at £100 each; they were required to establish not only bloomeries to reduce the ore, but forges where the iron could be hammered and wrought into bar stock.[2] Their first attempts to manufacture iron were made at Lynn, on the banks of the Saugus River; a second works was established at Braintree in 1648. Many bloomeries were in operation throughout New England and the Middle Atlantic colonies in the seventeenth century, and imports of both bar iron and iron finished goods decreased. In 1740 the British House of Commons held hearings on the reduced use of bar iron in England, which was in part the result of this diminishing colonial market for finished products. During the hearings it was stated that since 1722 no English axes had been shipped to New England, New York, or Pennsylvania, and only a few to the Carolina provinces. In his testimony, the English ironmaster William Carey said that iron of quality equal to the Swedish product was being made in colonial America.[3]

Although in the eighteenth century the blacksmith continued to make most American axes, some were made by early manufacturing tool-makers. Of the latter, one of the more suc-

FIGURE 22. American broad axe. Tool for squaring timbers, made from cast steel and marked "R. King." Length of blade, 12⅞ inches.

FIGURE 23. Pennsylvania broad axe. The term "goosewing" is popularly applied to axes of this pattern. Maker: Beatty, in Chester area of Pennsylvania, ca. 1875. Length of blade, 11¾ inches.

cessful and inventive was Hugh Orr. A Scotsman, proficient in blacksmithing, he came to New England in 1740. At East Bridgewater, Massachusetts, he worked for a scythe-maker and ultimately took over the shop. He is reputed to be the first American tool-maker to utilize a powered trip hammer in edge-tool manufacture. Through his efforts, axe and scythe manufactories were introduced in Rhode Island and Connecticut. A commentator on the early iron industry wrote, "For several years he was the only edge tool maker in this part of the country, and ship carpenters, mill-wrights, &c in this county [Plymouth], and state of Rhode Island, constantly resorted to him for supply."[4]

Orr was one of many who deliberately ignored British legislation on manufacturing. From the time of the Wool Act of 1699, the British Parliament fostered a mercantilist system in which the British colonies in the West Indies and America were to produce raw materials rather than finished goods. The Iron Act of 1750 was a further expression of mercantilist principles. It was intended to encourage iron smelting and the production of pig and bar iron in the colonial possessions, and to prohibit the establishment there of forges and slitting mills necessary for making tools, hardware, and other finished iron products. The Iron Act, however, had little effect on manufacturing in colonial America. Its enforcement was difficult because of the great distance between the seat of home government and the colonies, the lack of interest and the inability of the royal governors in carrying out its provisions, and the needs and ambitions of the colonists. There were, indeed, more forges in the American colonies at the advent of the Revolutionary War than in all of England and Wales.

FIGURE 24. Mortise axe. A blacksmith welded the 5¾-inch mortise bit to a worn-out felling axe. For rough work, such as cleaning holes in fence posts. Length of head, 10⅝ inches.

It is reasonably well established that the American axe-maker contributed a new form of felling axe in the eighteenth century. Until the introduction of the "Yankee" pattern axe, or "American axe," the poll, or portion of the head above the eye, had been minimal in size and weight. The new form had a large, heavy poll that outweighed the bit portion, and a shorter blade; this combination gave a critically different balance to the tool. In addition, the sides of the bit were formed with a wider, slightly rounded taper to prevent the axe being wedged in the tree on the chopping strokes. The conventional length of the handle was reduced. The form was innovative, yet like many modern tools it was but an up-dated form of the Stone Age implement. Comparing the "Yankee" felling axe with the Neolithic stone axe, Raphael Salaman notes that "both are smooth wedges with swollen sides so that the axe cleaves and cuts at the same time."[5]

The Duke de La Rochefoucauld-Liancourt was greatly impressed by the axes he saw while visiting America between 1795 and 1797. He noted that most of them had been made in America, and remarked: "We should not omit to observe, that the axe, of which the Americans make use for felling trees, has a shorter handle than that of European wood-cutters. Not only the Americans, but Irish and German workmen have assured me, that they can do more work with this short handled axe, than with the European. The blade is likewise not so large as that of the latter."[6] By "not so large," La Rochefoucauld was probably describing the length of the blade, because in the American pattern it was considerably shorter than in the English or European tool.

It is not at all clear when the new form of American axe became the type most generally used for land clearing and lumbering. The English observer William Douglass wrote in 1752 that "New England perhaps excels in good ax-men for felling of trees, and squaring of

FIGURE 25. Splitting and felling axes. *Left:* Holtzaxe (*Holzaxt*). Exclusively a Pennsylvania German eighteenth- and nineteenth-century splitting tool with heavy, wedge-shaped head. Length of head, 6¾ inches. *Right:* "Yankee"-pattern felling axe. Length of head, 7⅜ inches. Both tools are hand forged.

timber."[7] It was no doubt the implement rather than the man that was superior: merely transplanting a man from one side of the Atlantic to the other would hardly have made him a better axeman. The heavy-polled, short-bitted American axe was illustrated in a 1789 newspaper advertisement of William Perkins, a Philadelphia blacksmith. Such evidence, together with the recovery of new style axes from Revolutionary War sites, indicates that the innovative form was well established prior to 1790.

There were many small town centers of axe manufacturing in the nineteenth century, such as North Wayne and Oakland, Maine, but preeminent in the trade were East Douglas, Massachusetts, where the Douglas Axe Manufacturing Company was located, and Collinsville, Connecticut, site of the Collins Company. The Douglas firm had its origins in a blacksmithing business established by Joseph and Oliver Hunt in 1798. Oliver's son Warren continued operations until his plant was bought out and merged into the Douglas company in 1836. Only two years later, a hardware catalog listed an extensive line of the Douglas products, including two qualities of axes made in five different patterns, broad axes, hatchets, and shingling hatchets. The Douglas axe was a successful product and the firm prospered during the middle of the century. It exported extensively, shipping its edge tools to Australia and New Zealand, England, the West Indies, South America, and to Mediterranean countries. In 1865 the company was making axes in three grades, of which the top two were marked respectively "MFD BY W. HUNT" and "D. SHARP." At that time it reported production of its second-line Sharp axes at seven to ten thousand dozen a year.

The history of the Collins Company in Connecticut spans more than one hundred and fifty years, starting with its establishment in Hartford about 1823. In 1826 the plant was moved to Collinsville, where its buildings extended for almost half a mile along the Farmington River. Collins had the advantage in its early years of the management and inventive skills of Elisha K. Root. Root not only introduced efficient manufacturing processes but also developed new machines especially for edge-tool production. Further notable success was to come to Root at Colt's Patent Firearms Manufacturing Company, where he designed drop hammers for forging and a number of machines for firearms manufacturing processes. Like Douglas, the Collins Company had a large business, both domestic and foreign. Axes and machetes were major export items, shipped to South America and Australia. Collins also produced hatchets, adzes, cast-steel sledges, and stone hammers, but its chief item was the axe it advertised as "The Standard of the World." J. Leander Bishop, visiting the factory in the 1860s when Collins employed six hundred and fifty workmen, described "novel and original machinery unlike any other in use. One of their machines cuts up the Iron, gives it the shape and form of an Axe, and punches the eye ready to receive the helve, making a stronger eye than is made by welding in the usual way."[8] The Collins Company ended production at its huge Connecticut plant in 1966, the clang of its drop forges stilled as most of its hand tool products were supplanted by newer lumbering methods and machinery.

THE ADZ

Some American craftsmen, particularly ship-wrights and coopers, made extensive use of the adz. Although it was found in many house-holds, seventeenth- and eighteenth-century New England estate inventories indicate that the adz, or "adds," as it was frequently identified, was by no means as common as the ubiquitous axe. The adz has been manufactured in a wide variety of shapes and sizes. The blade is normally forged in a curve and is always at right angles to the handle. This allows the tool to be swung in an arc, to pare flat or curved surfaces and remove either thin or relatively heavy chips. The basil is on the inner side of the cutting edge. The handle of the American tool

may be short, as in the cooper's adz or small adzes used for hollowing bowls or chair seats; or it may be as long as thirty inches, as in the gutter adz used to hollow wooden rain gutters, or in carpenter's or in shipwright's adzes.

Most early adzes, like axes, were forged without a poll. A round eye was formed in the metal, into which the handle was inserted. As tool-makers developed adzes for the specialized trades of the carpenter, shipwright, and cooper, they added polls opposite the blade to serve discrete functions, such as hammering nails and spikes or pounding down hoops around the staves in assembling a barrel.

In the nineteenth century, manufacturers

FIGURE 26. Hand adzes. *Left:* Rudely made tool, the blade formed in a simple welded twist. Length of head, 7 inches. *Right:* Strap adz, a hand style used from antiquity to the twentieth century. Length of blade, 7 inches.

FIGURE 27. Cooper's adz. The short haft is convenient in barrel-making. Length of head, 8½ inches.

FIGURE 28. Shipwright's adz. The octagonal poll, almost 1 inch across the flats, is unusually large in this example. Marked "S. L. Cumins." Length of head, 10⅛ inches; of handle, 29 inches.

provided the adz with a novel form of socket for the handle. The head was formed with an extended socket which provided a longer eye. The eye was rectangular, rather than round, its sides tapering toward the handle of the tool. Thus the wooden handle, which flared slightly at its front end, had to be inserted from the front face of ·the head. The tapering socket effectively prevented the head from flying off the tool as it was swung.

In preparing timbers for house construction, the carpenter usually used the broad axe to square a post or beam smoothly, irrespective of whether these timbers were to remain exposed or to be boxed with planed boards. In rough work, however, such as barn construction, rapid squaring was often accomplished with the carpenter's adz. In both country house and barn construction, in fact, the common roof rafters were occasionally finished by the adz on the top surface only, the rest of the rafter left round, sometimes with the bark of the tree still intact.

Coopers used a short-handled adz for shaping the bevel on the inside stave ends at top and bottom; this bevel permitted the heads to be forced into place more easily. The poll end of the adz, a hammer head, was used in riveting and setting hoops.

The adz specifically designed for the shipwright was in constant use in framing vessels. The cutting edge of the blade was made from four to four and a half inches wide (twice the width of a cooper's adz). The usual poll of the shipwright's adz was a tapered pin over two inches long; it was used to drive the heads of large spikes below the surface of the wood. The shipwright might use the full length of the adz handle in shaping a mast, swinging the tool between his legs. But in shaping the frames of a ship for applying the planking, the tool was held with a shortened grip to cut at a chest-high level. The cutting edge of a variant design of shipwright's adz was made with lips at the sides of the blade; when used across the grain of the work, this conformation of the blade prevented the wood beyond the width of the chip being removed from splitting.

PERCUSSION TOOLS

The wood clubs that ancient carpenters and cabinetmakers used to strike chisels, gouges, and other cutting and splitting tools have survived to the present day in the form of carpenter's and wood-carver's mallets. A type common in early American industry was the froe club. This was customarily made from hard maple, beech, or other hardwoods. The maker simply sawed a heavy tree limb of about five inches in diameter to a length of fourteen to twenty inches, and reduced the diameter of one end to a convenient handle size with a drawing knife.

The handle of the iron claw hammer was first attached by fitting it into a round eye in the head, but during the Middle Ages, to strengthen the attachment of the handle, two iron straps were added which extended from the eye along the sides of the handle and were riveted through both head and handle. Later these straps were integrally forged with the head, and this method of hafting the tool is still used for some European and British hammers. Still another method of attaching the head, frequently used for American hand-forged hammers, was to run two straps along the handle

FIGURE 29. Claw hammers. *Left:* Straps are integrally forged with the head. *Center and right:* Straps are riveted to the handles, run through the eyes, and clinched. Length of heads, 3¾ to 4⅜ inches.

FIGURE 30.   Mallet. Rectangular head with gnarled and swirling grain is made from the burl of a tree. Length of head, 5½ inches.

and through the eye, and turn the projecting ends to prevent the head from becoming loose. David Maydole of Norwich, New York, who established his hammer manufactory in 1843, introduced the "adz-eye" hammer. He extended the eye for an inch or so, providing a firmer support for the handle; this has become the standard design of the American carpenter's hammer. The Maydole adz-eye hammer was offered in 1873 in head weights of seven and one-half ounces to twenty ounces.

While the hammer face has historically been of a relatively consistent shape — normally round, though occasionally rectangular or octagonal — the poll may take a variety of forms for specialized uses. The lathing hatchet, for example, which is a hammer on its striking end, has a long, knife-edged poll for rough chopping and fitting of wooden lath. (Lath strips were nailed to wall studs prior to plastering, and since the interior walls in early homes of the Plymouth Plantation period were not often plastered, it is somewhat surprising that a "lathing hamer" appears in the tool inventory of Godbert Godbertson of Plymouth, who died in 1633.)

The carpenter and shipwright used wooden mallets regularly. The cylindical wooden head of mallets and the heavier mauls was usually bound with iron rings at each end to prevent splitting. The heavy maul, known as a "beetle" or "commander," was made with a head six inches or more in diameter and at least a foot long. The carpenter used the commander to drive together the joints of house and barn timbers. "Beetle rings" are very frequently listed in estate inventories, which gives us some idea of the value the colonists placed on iron and iron implements.

Shipwrights used beetles and mallets for fitting structural members of framing and for driving treenails. By the nineteenth century another hammer form, the iron pin maul, had replaced the wooden maul in shipyards. The poll of this maul was formed in an extended, tapered pin. The shipwright used the hammer face of the maul to drive wooden treenails and iron spikes, and he used the pin poll just as he used the pin poll of his adz, to set the heads of the spikes below the surface of the planking. A specialized craftsman of the shipyard, the caulker, forced tarred hemp caulking material into the seams of planking by striking an iron caulking tool with another hammer form, the caulking mallet. This was made with a fifteen-inch-long cylindrical head. To judge by those used in New England shipyards and offered for sale in a catalog of the 1870s, lignum vitae and southern live oak were preferred for the heads of these mallets.

THE CHISEL AND GOUGE

The first settlers of America possessed both chisels and gouges. Estate inventories are once again an invaluable source. Among the tools left by Mary Ring of Plymouth Plantation was one "chisell." Other members of the Plantation owned tools identified as "gowges," "a great gouge," and both "broade" and "narrow" chisels.[9] Among twelve tools left by Edward Veir of Wethersfield, Connecticut, were a "pricker & chessell," the former an awl for making holes in wood, leather, or cloth. Veir was a man of very modest means and without family, who worked as a carpenter and joiner as well as farmed. His land consisted of two acres of meadow and a small additional "peece of land in Pennywise" valued at £1 10s. The balance of

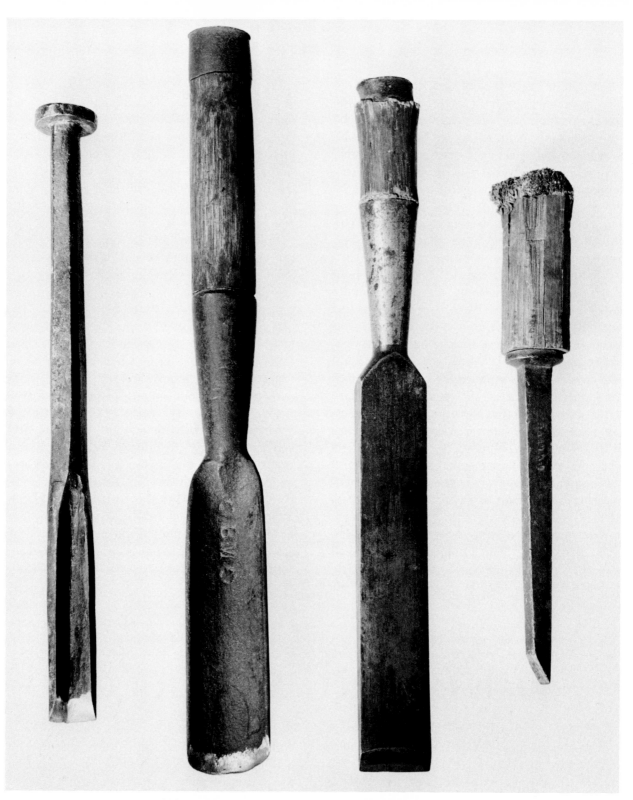

FIGURE 31. Chisel and gouge. *Left:* Hand-wrought corner chisel. Length, 12⅛ inches. *Center left:* Gouge, 1⅝-inch cut. Length, 15 inches. *Center right:* Socket framing chisel, 1½-inch cutting edge. Length, 14⅜ inches. *Right:* Mortise chisel. Length, 10⅜ inches.

his estate was a saw, a heifer, a barrel, an old brass pot, bedding, a lock and key, a bearskin, an old chest, a frying pan, other tools, and sums due him for corn, peas, and wheat, and a few small debts. We know that he placed a special value on his woodworking tools, for in the concluding bequest in his will of July 19, 1645, he states: "I give to Mr. Swayne all my workeing tooles."[10]

The chisel was important for the carpenter, shipwright, joiner, cabinetmaker, wheelwright, and millwright. There are two basic forms: the firmer chisel and the paring chisel, both made since dynastic Egypt. Both types were made in various sizes and lengths with a blade that is rectangular in cross section; the body of the firmer was relatively heavier than that of the paring tool. The firmer is struck with a mallet, while the paring chisel cuts as it is pushed. Two methods have been used to attach a wooden handle to either firmer or paring tool: the socket and the tang. The socket chisel has a conical socket above the blade to receive a handle with a round taper on its lower end; the tanged chisel terminates above the blade in a shank with an elongated spur that is inserted into the handle. In spite of the considerable antiquity of both socket and tanged chisels, the early American blacksmith-made tool was often a solid iron implement whose extended shank formed the handle.

The factories that specialized in the manufacture of chisels and gouges produced a staggering variety of these tools. The 1890 catalog of one of the foremost manufacturers, Buck Brothers of Millbury, Massachusetts, contains eighteen pages of specifications for tang chisels and gouges and an equal number for the socket variety. A socket paring chisel with beveled back and eight-inch-long blade was made by Buck Brothers in cutting widths from one-eighth inch to one inch, in eighth-inch increments, and from one to two inches in quarter-inch increments. Ten varieties of sized sets were offered, up to twelve chisels in a set.

The millwright's firmer chisel was a heavy implement for shaping mortises and tenons in timber construction. The D. R. Barton firm in Rochester, New York, made this tool with a twelve-inch blade in sizes up to four inches wide. A four-inch millwright's socket firmer chisel with a twelve-inch blade, twenty inches long overall without its handle, weighed over nine pounds. This was obviously a two-man tool: one man positioned and held it for a cut, and a second man drove it with a heavy maul. The carpenter's and shipwright's slick was another large form of chisel, made with cutting blades two and one-half to four inches wide. It had a lathe-turned wooden handle twelve to fifteen inches long, which terminated at the upper end in a boss. The tool was hand held, pushed rather than struck, and used for fine paring of wide surfaces, the flat sides of large tenons, or the insides of mortises. Some carpenters worked the tool by pressure from the shoulder, using two hands to guide the path of the blade. The corner chisel was one of many special-purpose chisels, a form of the firmer chisel with blade shaped in a right angle to cut simultaneously in two planes. For clearing out and squaring the corners of mortises, it was fitted with a handle made to be struck with a mallet.

The gouge, whose blade forms an arc in cross section, produced a rounded cut. Gouges were made in three dimensions of curvature: full sweep, middle sweep, and flat sweep, to produce a deep, a medium, or a shallow concave cut.

Many nineteenth-century edge-tool firms with a diversified line manufactured chisels and gouges. The Underhill Edge Tool Company of Nashua, New Hampshire, which was established shortly after the Revolutionary War; Buck Brothers; D. R. Barton; L. & I. J. White of Buffalo, New York; the James Swan Company of Seymour, Connecticut; and Peck, Stow & Wilcox of Southington, Connecticut, were all major manufacturers.

THE SAW

At the time of the colonization of America the saw was refined in design, if not in materials or method of manufacture. It was made in many forms: the hand saw; the small frame saw with blade held in tension in either a rectangular or bow-shaped wooden frame; the thwart saw, a two-handled crosscutting style for sawing lumber and logs; the large five- to ten-foot, two-man pit saw, which was a rip saw for cutting large timbers, the saw either held in a large wooden frame, or "open," with a bare blade and handles at either end; and saws in many other special-purpose sizes and styles (Figure 80). The two main types are the crosscut, used to cut across the grain of the wood, and the rip saw, used to cut along the length of the grain. The crosscut hand saw has from five to eleven points to the inch and teeth raked toward the point of the blade. The hand rip saw normally has from four and one-half to six points to the inch. The teeth are larger than those of the crosscut, and they are filed with the front cutting edge approximately perpendicular to the length of the saw. Since the latter part of the nineteenth century, the rip saw has been manufactured with a graduated number of points, there being one more point to the inch at the front end of the saw than at the butt, or handle, end. Measurement of the points of crosscut and rip saws is made at the butt end of the blade, and there is always one less tooth to the inch than the number of the points. The teeth of both crosscut and rip saws are set to right and left. Setting saw teeth properly was an occupation that demanded a high degree of skill and a sensitive touch with the setting hammer. Every other tooth was struck on a small anvil, the blade was turned over, and the alternate teeth were bent in the opposite direction. The artisan had to make the set of all teeth equal so that the blade would cut smoothly. In the nineteenth

century a number of devices were patented and manufactured for setting the teeth mechanically (Figure 34).

American saw manufacturing did not come into its own until the second half of the nineteenth century; prior to that time most saws were imported. England was the major source, and the superior products of the Sheffield saw firms established a continuing reputation for quality. It was not until midcentury that the primary processes of iron manufacture in America began to catch up to the technologically advanced production of Great Britain and some of the European countries. Then the substitution of mineral fuels for charcoal, the use of a hot blast furnace in place of a bellows operated by waterpower, the wide application of rolling in the refining process, and a better understanding of the properties of ores all contributed to the manufacture of better grades of steel. For saws and other tools requiring hard and flexible steel it was no longer necessary to rely on importation.

Some saws, however, had been manufactured in America at scattered locations as early as the pre-Revolutionary period. They were among the tools being made in the late years of the eighteenth century in Plymouth and Bristol counties of Massachusetts, where furnaces, rolling mills, and slitting mills were clustered. It was not until the early 1800s, however, that there was any sizable production of saws. In 1802 William Rowland established a shop for making saws in Philadelphia and the firm continued in operation until 1851.

Charles Griffiths, who had gained experience in England, came to Boston in 1830 and established a saw factory in West Cambridge, Massachusetts. The firm, later known as Welch and Griffiths, successfully produced a full line of saws, from hand saws to large circular and up-

FIGURE 32.   Stair saw. Usually made by the saw manufacturer, this example was forged by a blacksmith from a scythe blade. Length, 9½ inches.

FIGURE 33.   Bow, or turning, saw. A form of framed saw; the blade is tensioned by twisting the top cord with the toggle stick. Handles can be turned to alter the angle of the blade in sawing curves. Length, 28¾ inches.

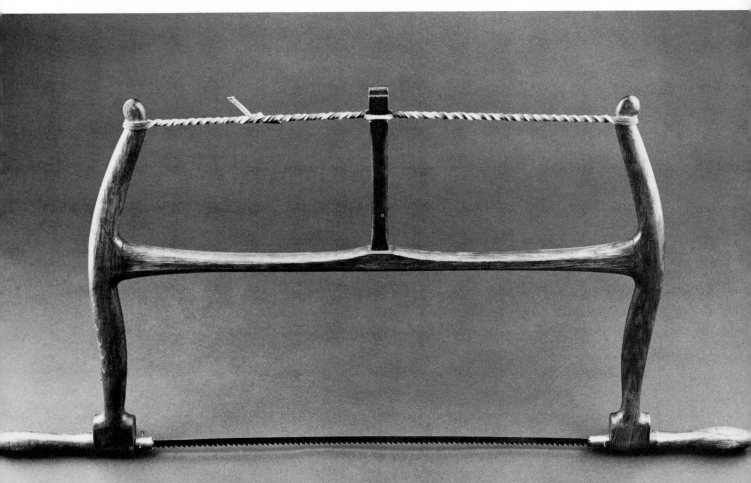

FIGURE 34. Herrick Aiken's saw set. Simple, yet effective mechanical device for setting hand saw teeth. Patented in 1830, it had a market for over one hundred years.

and-down mill saws. J. Leander Bishop, writing in 1868, stated that much credit for overcoming American prejudice against locally manufactured saws could be given to Welch and Griffiths. The quality of their product was such that it successfully met the competition of imported saws.[11] Henry Disston, like Griffiths an emigrant from England, solidly established America as a major saw producer, both for the home market and for worldwide export. The Philadelphia plant of Disston's Keystone Saw, Tool, Steel, and File Works, which was established in 1840 and ultimately spread over fifty acres, included all phases of manufacture from raw materials to finished products. Steel for making the saws was produced in its rolling mills, and a woodworking department made handles for the saws as well as for bevels, squares, and other tools. Although the Disston factory produced all the steel used in its handsaw blades, it identified that used in several of its best and most expensive saws as "Extra Refined London Spring Steel," probably as a gesture toward the lingering reputation of the English product. It qualified this English attribution (which was merely advertising verbiage) by also stating that these saws were made from "Disston Highest Quality Crucible Steel."[12]

Many saw factories were started in the middle of the nineteenth century. Hasbrouck's *Middletown Directory for 1857–'8* recorded the establishment of the Wheeler, Madden & Bakewell firm in that New York State community in 1853. It averred that Josiah Bakewell, practical mechanic, "is reported to be the best saw maker in the United States," and his company's product "equal both in essential quality and style of finish to any in the market, whether of English or American production." Superlatives appear to have been the order of the day in saw advertising. Emerson, Smith & Co. of Beaver Falls, Pennsylvania, makers of "$100 Gold Premium Saws," characterized their product as "Damascus Tempered." R. Hoe & Company of New York City advertised their saws made from cast steel as "far superior in every respect to any others in the market." Saw-making was carried on in New England by small firms such as the Fisherville Saw Co. of Penacook, New Hampshire, and by large ones such as the Simonds Manufacturing Co. of Fitchburg, Massachusetts, which established branch plants throughout the country and in Canada. As population and industry moved westward, saw companies were established in the midwest; E. C. Atkins & Co. of Indianapolis was an important manufacturer of hand saws for the carpenter and

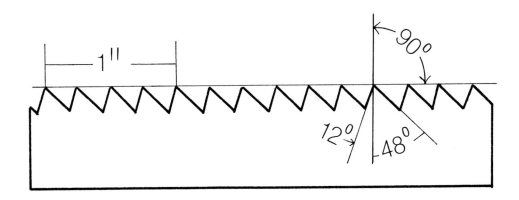

FIGURE 35. Crosscut saw teeth. Angle of rake, 12 degrees off vertical; used by Disston ca. 1900. A five-point saw has four teeth to the inch.

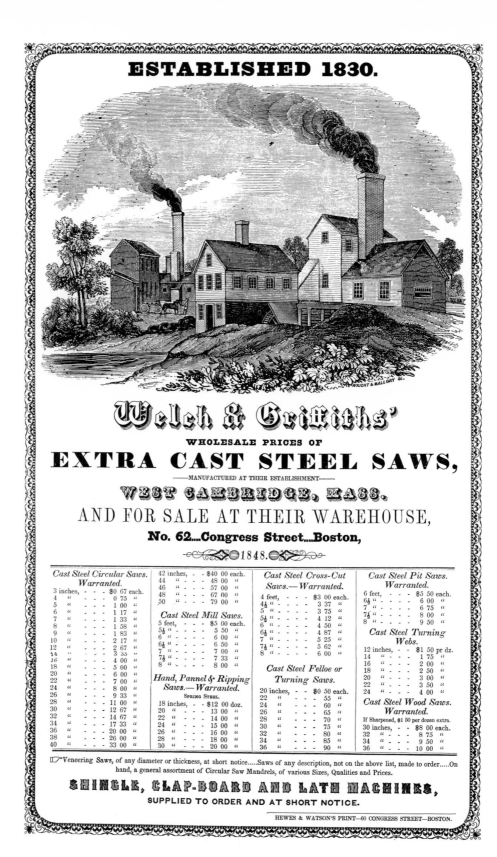

FIGURE 36. Advertising broadside and price list, 1848.

Disston & Sons' Cross Cut Saws.

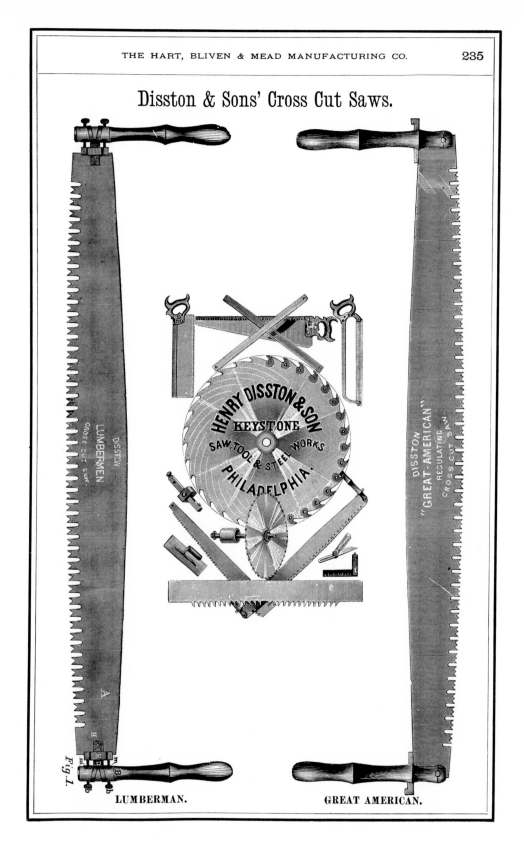

LUMBERMAN.   GREAT AMERICAN.

FIGURE 37.   Disston saw advertisement in tool and hardware catalog of 1873.

53

large saws for the lumbering and sawmill industries. By 1860, there were forty-two establishments, operating in eleven states, whose only or chief product was saws. They employed over seven hundred and fifty workmen, and the value of their annual production was more than one million, two hundred thousand dollars. These census figures were probably conservative.

BORING IMPLEMENTS

The bow drill used in America, even into the twentieth century, was functionally similar to the ancient tool but made of different materials. The cap the Egyptian artisan had held in the palm of his hand became an integral, swiveling head for the shaft; steel formed the rotating shaft to hold the bit; and brass, ivory, and ebony were often utilized for both functional and decorative parts. The bow was made from steel, with a turned handle, often of rosewood. In America the woodworker's bow drill became associated particularly with piano-making. It served well for drilling the many holes in the piano wrest-plank, into which tuning pins were inserted. Stone-carvers also used the bow drill for surface cutting of granite and marble; for this, the bow string was of brass or steel wire, wound on gut.

Although the bow drill was developed earlier, the auger has been the most common form of boring tool for centuries. A unique Egyptian specimen, from the Twenty-sixth Dynasty, about 550 B.C.,[13] has been found; there are Roman and Viking examples; and the auger is depicted in several medieval book and manuscript illustrations. The tool has traditionally been made with a forged iron shaft that terminates in variously shaped cutting surfaces. A transverse wooden handle at the top, which the workman rotated and simultaneously pushed, produced the boring action. Roman and medieval augers are generally of a spoon pattern, but in the tenth century a short spiral twist, effective in removing chips as the tool was turned and forced through the wood, was also developed. This design, which again became a standard feature of most augers of the 1800s, appears to have been abandoned by the seventeenth century (or possibly even earlier) in favor of a semicircular, half-cylinder pattern, terminating in an almost horizontal cutting lip. The augers used in the early days of America were of this latter design and were identified as pod auger, nose auger, shell auger, or spoon auger, depending on the shape of the terminal cutter (Figure 40). Lacking a center point, the nose auger was particularly difficult to start in wood, and a preliminary starting hole or indentation with a gouge was required. Although some early-nineteenth-century semicylindrical augers were provided with a screw point, an important aid in drawing the tool through the wood, the lead screw device came into general use only with the development of the spiral auger.

There are two general classes of spiral auger, the single twist and the double twist. The spiral of the former is made by twisting an iron rod of roughly triangular shape. The double twist is formed by twisting a relatively flat metal blank, creating two continuous helixes (Figure 39). Like the plow, the spiral auger was a favorite subject for the ingenuity of American inventors. A variety of designs for the shearing edges of the horizontal cutters and their accompanying

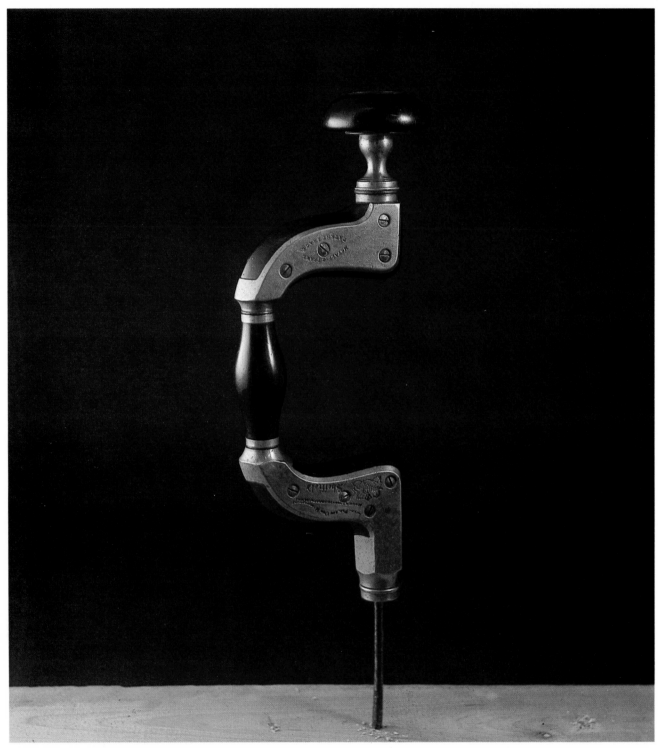

PLATE 1. (*Overleaf*) Bit stocks. *Left:* Sheffield brass-plated brace. Length, 14 inches. *Center:* Well-proportioned homemade tool with clothespin type bit pad. Length, without pad, 15⅜ inches. *Right:* As manufactured in New York State by G. Benjamin. Length, 15 inches.

PLATE 2. The "Ne Plus Ultra Framed Brace." Metallic frame patent brace, in cast brass and ebony, of "Henry Pasley's Own Manufacture," from Sheffield, England. Length, 12¹⁵⁄₁₆ inches.

PLATE 3. For determining a true vertical, the suspended plumb bob of brass or iron.

PLATE 4. Plow plane with sliding arms. By Robert Wooding, who worked in London from 1704 to 1728. A prototype of the tool as made by eighteenth-century American plane-makers. Length, 8⁷⁄₁₆ inches.

PLATE 5. Chalk-line reel. Fashioned of bent ash wood. Length, 12½ inches.

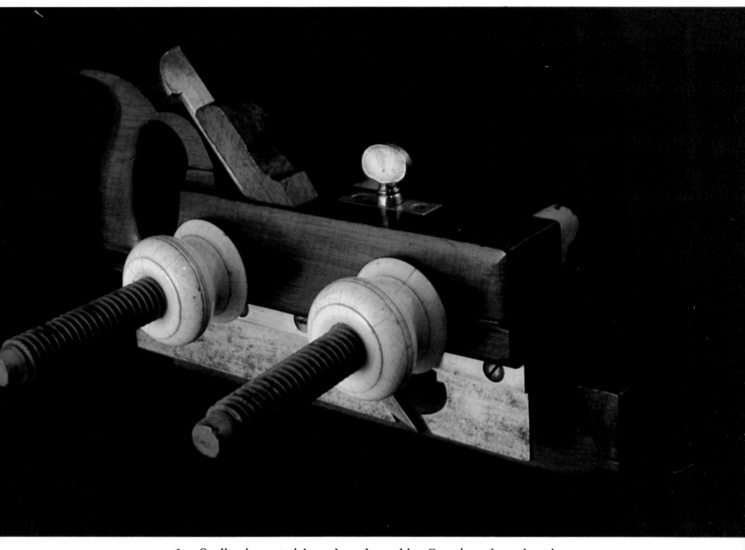

PLATE 6.   Quality in materials and workmanship. Grooving plow plane in rosewood with threaded arms of boxwood. The stop nuts for adjustment of the fence are
of ivory. Made by Greenfield Tool Co., Greenfield, Massachusetts. Length, 12 inches.

PLATE 7.   Hand saw handles. Artistry of shape in saw handles of tools made by Henry Disston and Josiah Bakewell about 1865.

PLATE 8.   For shipbuilding. *Above:* Slick, an oversize paring chisel. Made by L. & I. J. White of Buffalo, New York, with 3-inch-wide cutting edge. Length, 27⅜ inches. *Below:* In the heyday of the wooden ship, some masts were turned on huge lathes, but a broad axe and a mast knife like this one were the usual tools for the job. Length, 29¼ inches; blade, 20¾ inches.

PLATE 9.   Shipwrights' planes. The products of ships' joiners. *Above:* Rosewood jointer. Length, 24 inches. *Center:* Fore, or short, jointer, made of a dense tropical wood. Length, 21⅛ inches. *Below:* Jack plane in lignum vitae. Length, 15⅞ inches.

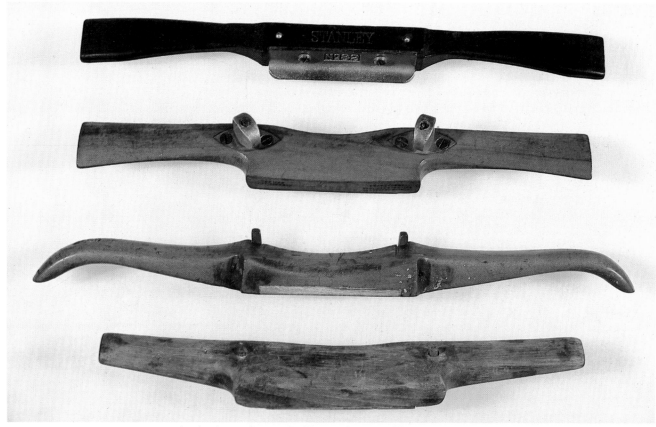

PLATE 10.   Spokeshaves in wood. *Top to bottom:* "Razor edge" shave by Stanley; thumbscrew adjustable model by John Booth & Sons, Philadelphia; and two shop-made shaves. Lengths: 10⅜ to 14 inches.

PLATE 11.   Coach-maker's router. The handles and stock are joined with mortises and tenons, through-pegged. Of birdseye maple. Length, 15¼ inches.

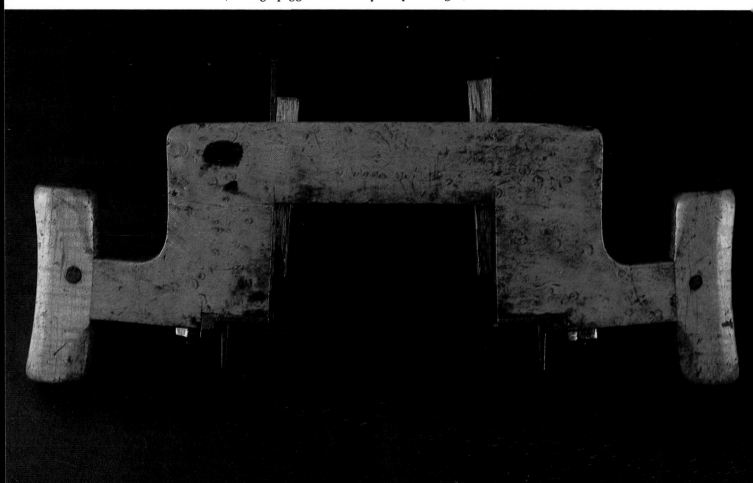

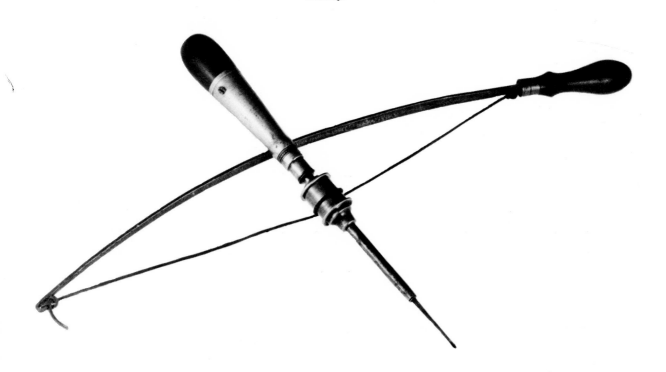

FIGURE 38. Bow drill. Reciprocal motion of the bow creates rapid rotation and counterrotation of the bit. Bow length, 23⅝ inches. Drill length, with bit removed, 12⅞ inches.

cutting spurs were patented. Manufacturers provided single- and double-thread screw points to pull the tool straight through the wood. Ezra L'Hommedieu of Chester, Connecticut, is credited with the basic design of the spiral auger, on the basis of an 1809 patent. His original patent specifications were destroyed in a fire at the United States Patent Office, but in a subsequent reconstruction of the patent L'Hommedieu indicated that two sharpened, vertical lips created the cutting action. In later patent designs of double-twist augers, the principal cutting was performed by the shearing action of two horizontal edges formed at the terminals of the spiral. An important feature of both single- and double-twist augers was the pas-

sage afforded by the continuous spiral for the automatic lifting and removal of the wood chips. Shell augers, on the other hand, made a hole by the action of the one horizontal cutter at the base of the hole, and the sharpened vertical edge scraped the circumference of the hole; these tools had to be repeatedly removed from the hole so that the chips would not clog the rotary motion. The majority of the inventions to improve the function of the spiral auger were related to design of the cutting lips and circumferential spurs, which define the size of the hole, or to chip removal. A typical patent was that of 1847 granted to Sanford, Newton, and Smith on construction of single- or double-twist augers that provided for "a gradually in-

FIGURE 39.   Auger-bit construction. From drop-forged steel blank, through the dies of a twisting machine, to finished double-twist bit with lead screw.

FIGURE 40. Patterns in augers. *Left to right:* Shell, or nose, auger; pod, a shape common in augers ten to twelve feet long for boring wooden water pipes; single twist; single twist with lead screw, made by L'Hommedieu; and double twist. Diameters, ¾ inch to 1⅛ inches.

FIGURE 41.   T-handle auger. A 3-inch-diameter tool, requiring great strength to operate. Length, 24 inches.

FIGURE 42. Sanford's incremental twist pattern. Bit patented in 1847, designed to improve removal of chips by a gradual increase in the cavity of the helix.

creasing length of twist" to facilitate discharge of the chips. This concept was retained for many years as one popular design for augers, and for the smaller auger bits used in bit braces.

The carpenter's auger was forged in lengths of from one to two feet. Cutting diameters ran from one-quarter inch up to as much as three and a half inches. The top of the shank was usually made in one of three different styles for affixing the transverse handle: a simple flattened tang onto which the handle was driven and the point of the tang clinched; a threaded tang, which was put through a hole in the center of the handle and capped with a nut; and an integral iron ring through which the handle was inserted.

The auger was essential to the shipwright. Every one of the treenails in a wooden ship required a hole that had to be bored by hand through oak planking and into oak frames. Nineteenth-century shipwrights favored the single-twist, cast-steel auger, to which, for convenience, they often added an elongated shank with a crank handle.

Another basic form, the tapered auger, was used by the cooper to bore the bung and tap holes in barrels. In wagon- and carriage-making, the wheelwright used a larger version to bore the tapered hole in a wheel hub. This hole accommodated the "box," a hollow iron cone that served as a bushing for the spindle of the axle.

Nineteenth-century Connecticut became a focal point for the manufacture of augers and auger bits, and for other drilling and boring tools. As Carlson and Stevens have pointed out, of some sixty patents on the cutting action of augers between 1790 and 1873, twenty-five were granted to Connecticut inventors. During this same period, twenty-one of twenty-nine patents on the machinery and processes for

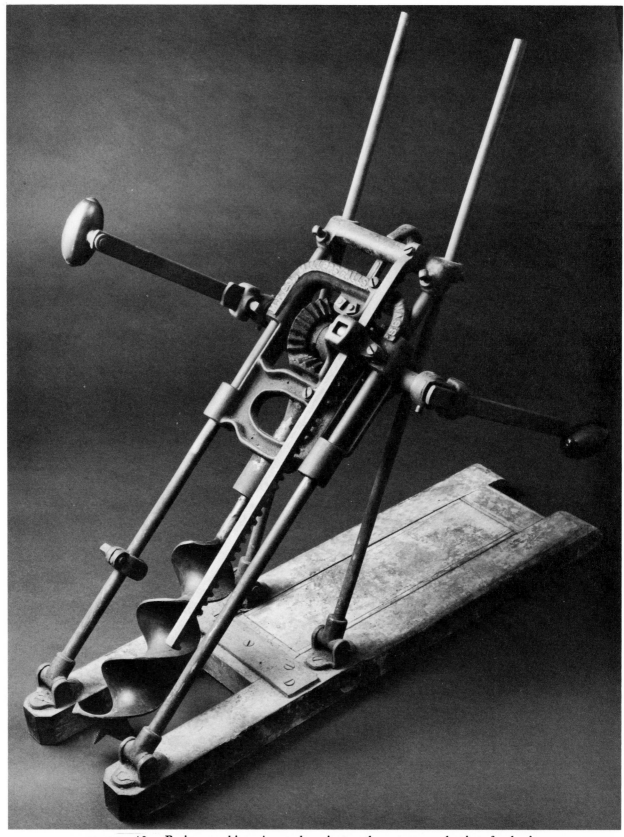

FIGURE 43.   Boring machine. A popular nineteenth-century mechanism for boring treenail holes in beams of house and barn construction. This model, made by the Millers Falls Co. of Massachusetts, can be adjusted for boring at an angle. Height, 25 inches; length of base, 27½ inches.

FIGURE 44. Jennings auger bits. Boxed set with three trays containing thirteen bits for use in a brace. Sized 4 to 16, ¼ inch to 1 inch, graduated in sixteenths of an inch.

manufacturing augers were also issued to residents of Connecticut.[14]

The earliest known representation of the bit brace is in the famous early-fifteenth-century triptych by the Flemish artist Robert Campin, known as the Master of Flémalle (in the Metropolitan Museum of Art, New York). In the right-hand panel, Joseph the Carpenter is depicted using a brace to drill holes in a rectangular piece of wood. The brace he holds is identical in design to those made with a wooden stock on this side of the Atlantic up to four hundred years later (Plate 1). The tool has

been known by a variety of names, including bit stock, brace, piercer, and wimble. In inventories of 1633, settlers in the Plymouth Plantation referred to the tool as a "piercer" or "piercer stock," while in Connecticut records it is listed as a "wimbell" in 1641 and 1643, and as a "brest wimble" in 1645.

The bit brace is essentially a toolholder for a bit. It consists of a round head, or breast block, which rotates freely on the stock; a main bow, or crank; and a chuck, or "pad," which is the bit holder. The brace was efficient for boring small holes because it provided a continuous rotary

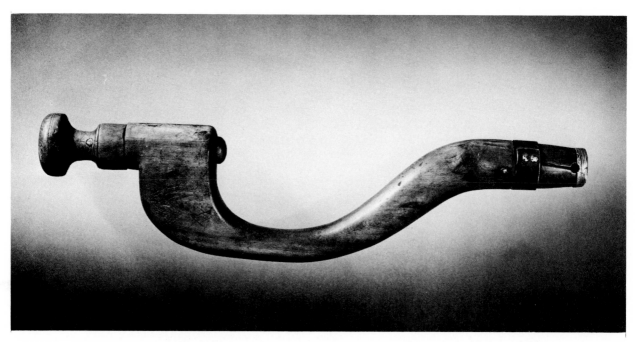

FIGURE 45.   Primitive bit stock. Large, handmade version of the tool. Length, 22⅞ inches; sweep, 8 inches.

motion for the cutting bit, rather than the intermittent action the auger provided. (With the latter the user made a half turn of the tool, then had to remove his hands to take a second half turn, and so on.) The "sweep" of a wooden brace, the diameter of a circle made by the rotation of the crank-handled stock, was small and customarily not more than seven or eight inches. An excessively wide sweep would allow greater force to be applied, but with the danger of fracturing the wooden stock. In the early Continental and American brace, the pad was not integral with the body of the tool; for each size hole to be bored, a separate, interchangeable wooden pad held a different-sized bit. The bit stock of this type was commonly made of beech, maple, oak, or other hardwood, and was usually a homemade rather than a manufactured product.

By the mid-nineteenth century the brace used by American joiners and cabinetmakers was made either of wood or iron. The factory-made tool with a wooden stock was frequently imported from England, from the tool-making cities of Sheffield and Birmingham. Although it was identified in at least one English tool catalog as a "Scotch" brace, it became better known as a "Sheffield" brace through association with its principal place of manufacture. Brown & Flather, Tillotson & Co., E. Windles, Slater, Dixon's Hermitage Works, and William Marples, all of Sheffield, produced a very similar product. A durable head of hard lignum vitae wood was fitted with a brass quill to the main stock. Beech wood was used for making the stock of the standard line. In the more expensive models the crank was reinforced by four cast-brass plates recessed into the surface of the wood at the bows of the crank. A cast-brass chuck accepted rectangular, tapered-shank bits. The chuck was provided with an internal spring catch, operated by a button. It was necessary to file a notch in the shank end of each bit; a projecting point on the internal spring catch en-

FIGURE 46.   Bit stock. A craftsman copied the Sheffield pattern. Pads with separate bits fit a pewter-bound socket. Round head is shaped from a burl. Length, 15½ inches; sweep, 6 inches.

gaged the notch and held the bit securely when inserted. Some of the most spectacular examples of the Sheffield tools made by William Marples and others were the patent metallic frame braces (Plate 2). Rather than wood with metal supporting parts, these were made of heavy brass with wood inserted into the metal framework. Ebony and rosewood were the usual choices for these expensive and highly finished items. Some came to America by importation in the nineteenth century, and others have followed only recently, in the past decade, to satisfy the American tool collectors' market.

At least one American firm was recorded as making the Sheffield-type brace. Booth & Mills of Philadelphia received awards from the Franklin Institute in 1856 and the American Institute of New York in 1857 for the quality of their hand tools, including bit stocks which were cited as "certainly noteworthy for their elegance." Their display at the 1876 Centennial Exhibition included a fine ebony brace with polished steel reinforcements in the style of the English tool. (In 1976 the Smithsonian Institution's National Museum of History and Technology, reconstituting that centennial exhibition, displayed the Booth & Mills exhibit case with this tool.) In 1857 George Benjamin of Avoca, New York, received a patent for "Improved Devices for Holding Bit in the Brace." One example of this patent, applied to a brass-plated bit stock with cast-iron chuck that approximates the Sheffield pattern, is stamped with Benjamin's name and the patent date no less than six times (Figure 49).

The metal bit stock, often a blacksmith-made tool, developed as a manufactured product in America during the nineteenth century. Among the tools carried by the William H. Carr firm of Philadelphia in 1838 were "Bit Stocks, Iron, Patent" made by Increase Wilson of New London, Connecticut; these were all-metal braces made on the patent of J. Taylor of Hebron, Connecticut, issued June 30, 1836. The Carr catalog, entitled *American Manufac-*

FIGURE 47. Six bits. *Above, left to right:* Center bit; bit stock drill; gimlet, or "German," bit; lip bit; pod, or gouge, bit. *Below:* Spoon bit, forged from an old file. Diameters, ¼ inch to 1¾ inches.

tured Hardware, &c., lists both iron braces and "Fine Brass Plated Braces" among the tools of S. C. Bemis & Co. If indeed Bemis did manufacture a wooden brace with brass plates, none are known to have survived, and it is more likely that in 1838 these were imported from Sheffield.

A ten-inch sweep was the standard size of the carpenter's metal bit brace. To accommodate requirements of different use, they were manufactured with sweeps of from six to sixteen inches, in increments of two inches. Nickle-plated metal stocks, with heads and crank handles of rosewood, lignum vitae, or cocobolo wood, gave the tools a finish that was both attractive and utilitarian (Figure 52). For top-of-the-line models, the makers provided ball bearing chucks and heads for ease of operation and reduction in wear at critical points, and completely enclosed ratchet actions.

On November 1, 1859, N. Spofford of Haverhill, Massachusetts, patented an improved bit stock. The patent specification described a new method of holding bits so that rectangular shanks of different sizes, with different degrees of taper, could be effectively accommodated. The design of this device was simplicity itself: a split in the front arm of the brace divided the socket into two parts; these were drawn together by a thumb screw to hold the bit shank.[15] Spofford's patent brace was a practical, inexpensive, and popular tool. It was manufactured well into the twentieth century, originally by the John Fray Company of Bridgeport, Connecticut, and ultimately by the Stanley Rule & Level Company; it was carried in Stanley's catalog of tools for carpenters and mechanics as late as 1941.

Perhaps the most important contribution to the manufacture of the American bit brace was that made by William Henry Barber of Greenfield, Massachusetts. The original patent for the Barber brace was issued on May 24, 1864.[16] It provided the basic design for a bit-holding chuck which has remained fundamentally unchanged for over one hundred years. A tubular

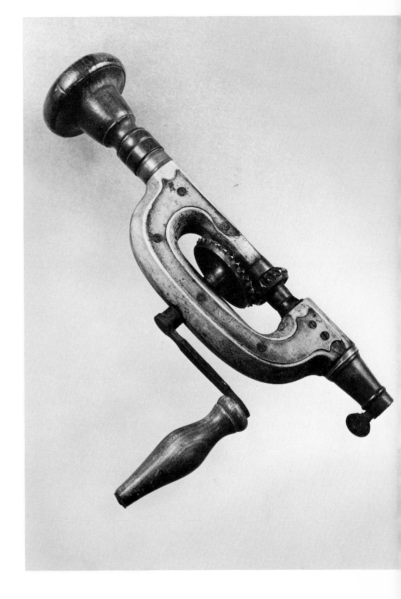

FIGURE 48. Hand drill. German tool, with inset brass reinforcing plates in the style of a Sheffield plated brace. Brass bevel gearing and chuck, with lignum vitae head. Length, 12¾ inches.

FIGURE 49. Imported and domestic bit braces. *Above:* An English plated brace by Tillotson & Co., Columbia Place, Sheffield. Length, 14 inches. *Below:* George Benjamin's American product. Length, 15 inches.

FIGURE 50. Patent bit brace chuck. George Benjamin's spring-loaded catch was a modification of an internal fitting in the Sheffield style of the tool.

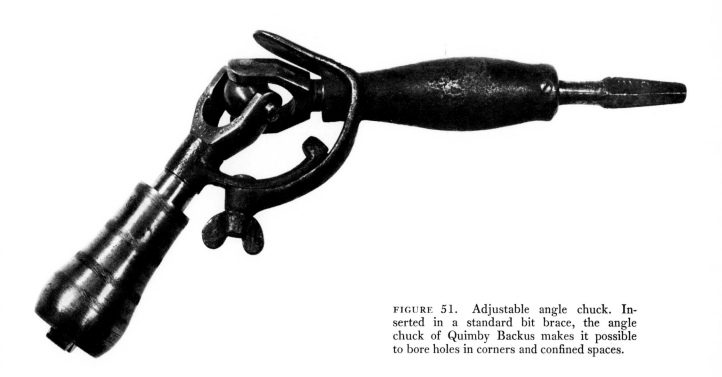

FIGURE 51. Adjustable angle chuck. Inserted in a standard bit brace, the angle chuck of Quimby Backus makes it possible to bore holes in corners and confined spaces.

outer sleeve, internally threaded, screwed onto a cylindrical threaded tool stock. A slot in this stock held a pair of serrated jaws. As the round sleeve was screwed onto the stock, it forced the jaws to grip firmly the bit's shank (Figure 53).

The manufacture of bit braces in the last quarter of the nineteenth century was centered in Connecticut and Massachusetts. A number of patents were issued for improvements in the tool; Greenleaf Stackpole, H. S. Bartholomew, Benjamin Darling, and Gardner S. Holt, among others, patented innovative chucks to hold the bit (Figure 54). A major refinement was the provision of a reversible ratchet to permit intermittent, partial movement of the crank where working space would not allow full and continuous rotation. One of the earliest ratchet devices, patented in 1857, was the invention of Henry Porter of Rothsville, Pennsylvania. The ratchet was added to the Barber brace made by the Millers Falls Company of Massachusetts before 1882. The Peck, Stow & Wilcox line of braces incorporated a ratchet device made on a Decem-

ber 30, 1884, patent, and the Davis Level and Tool Company of Springfield, Massachusetts, introduced John Bolen's ratchet patent in 1886. The ratchet provided flexibility, but with an understandable sacrifice in speed of operation. To make possible continuous boring "near the wall or other fixed object which would interfere with the use of an ordinary bit stock" (in the words of the patent), an angular bit holder with a double universal joint was developed by John F. Cory of New York City and patented on June 25, 1861. Cory's adjustable-angle chuck was a modification of the universal-joint model patented in 1857 by Charles Plaisted of Chicopee, Massachusetts. Similar devices for angle boring were subsequently perfected, including one by Quimby S. Backus of Millers Falls, Massachusetts, patented in November 1872. This type of angle drive, with either universal-joint or bevel-gear transmission, was later incorporated in a bit stock called a "corner brace" designed for carpenters, electricians, and plumbers.

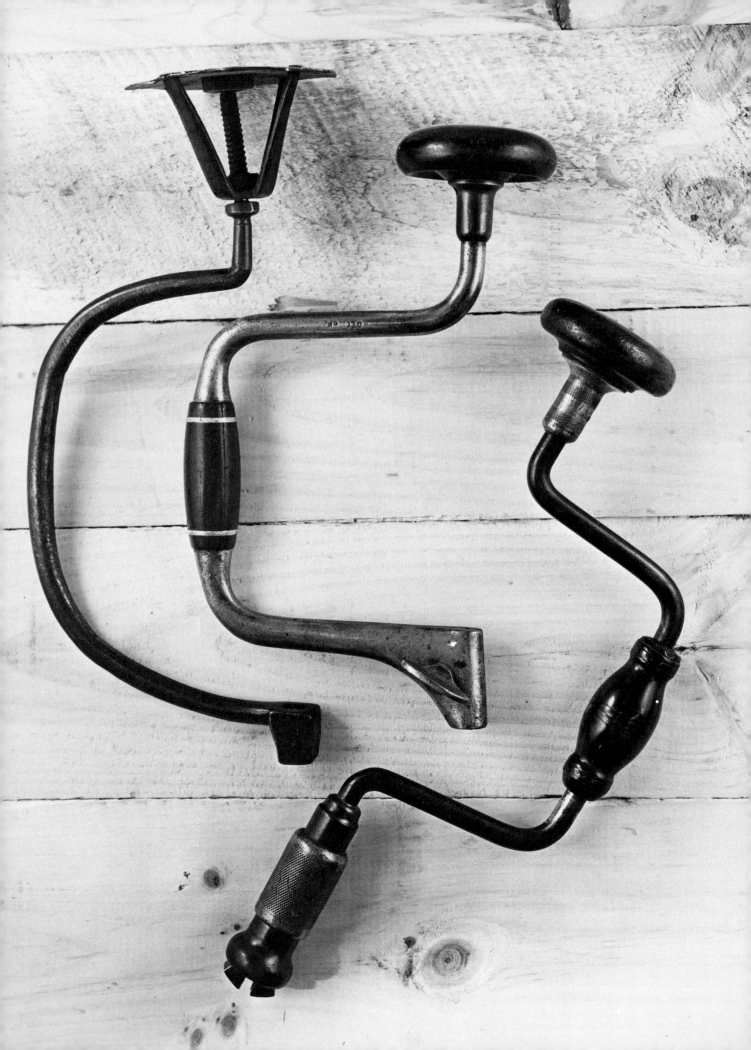

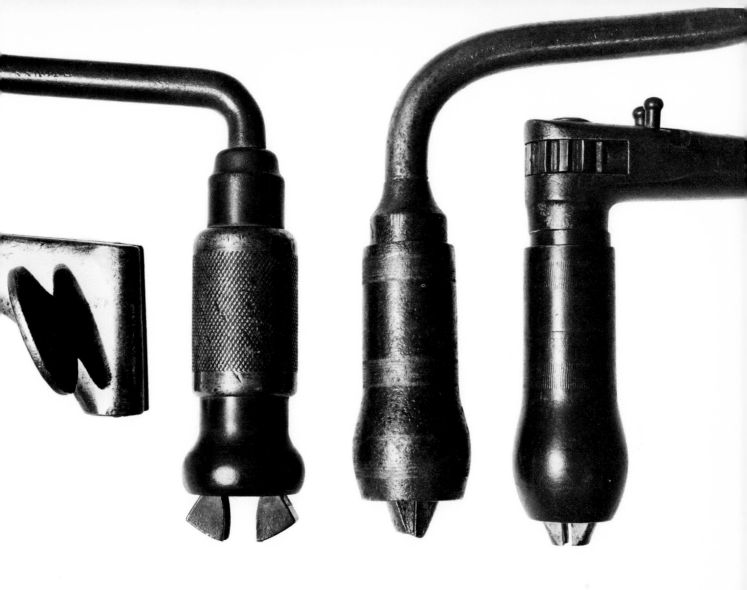

(*Opposite.*)

FIGURE 52. Metal replaces wood. *Left to right:* Hand-forged carriage-maker's cage-head brace; length, 12¾ inches. Spofford brace, patent of 1859; length 11 inches. Davis Level & Tool Co. brace with brass chuck and gutta-percha handle on the crank; length, 14 inches.

FIGURE 53. Innovation in chucks. *Left to right:* Spofford's split chuck; Davis chuck patented in 1883 and 1884; Barber's original shell chuck design of May 24, 1864; Peck, Stow & Wilcox chuck, with reversible ratchet patented by Amos Shepard of Plantsville, Connecticut, in 1884.

FIGURE 54. Split-socket bit holder. One of many ideas for gripping the bit shank, this one was invented by Greenleaf Stackpole of Portland, Maine, in 1862. All were ultimately supplanted by Barber's shell chuck.

THE PLANE

The tool that has survived in greatest numbers from the first three hundred years of hand woodworking in America is the plane. The importance of the plane lies in the flexibility of control it affords. Prior to its invention, smoothing, paring, and fitting of wood was of necessity done by a variety of methods and with a number of tools. The adz, in skilled hands, could be used to produce a relatively smooth surface but not one sufficiently precise for joining boards together with a close fit. Chisels could also serve for surfacing when used with a fine paring action. But it was not possible to control the angle and depth of cut, or to surface in a consistently straight line until the plane was developed. The usual form of the plane does simply what its name suggests: it creates a surface that wholly contains every straight line joining any two points thereon. The plane stock, in effect, is a holding implement for a chisel-like cutter, with a flat sole through which the cutter projects at an angle. It has a means to regulate the extension of the cutting iron below the working surface. This feature permits the user to remove irregularities of surface rapidly with a lower, or "rank," setting of the iron, or to perform finish planing with a fine setting of the iron.

For nineteen hundred years the plane remained relatively unchanged in design. Planes with completely flat soles for surfacing and edge jointing, molding planes to shape a variety of decorative surfaces, and grooving planes were in use from the Roman period, through the Middle Ages, and into modern times. Through much of the medieval and Renaissance periods, however, joiners and cabinetmakers ornamented wood surfaces by carving rather than by the use of molding planes. By the seventeenth century, because of changes in styles of cabinetry and interior decoration, molding planes as well as bench planes were used both in England and

America. The stocks were of wood; although the Romans had used metal in the body of the plane, the use of metal (other than for the cutting iron) was abandoned early. The plane chiefly or entirely of metal was not revived until the nineteenth century.

From the late Middle Ages to the eighteenth century the tools used in England were basically similar to those of Europe with one major exception — the plane. There is an unfortunate dearth of English planes from the period before 1700, and we can only infer from those of the first known English plane-makers (Figure 7) that their makers were carrying forward the earlier traditional styles with which they were familiar. European planes of this period, however, are well represented in museum collections. Characterized by decorative surface carving, they normally provide an upright front handle; called in small versions a "horn" plane because of the shape of this front handle, the style is still followed today by German and Austrian manufacturers. The larger bench planes generally were provided with handles at both front and rear. A few planes from the Continent and Scandinavia have survived that lack the usual extensive carving. It is very possible that many undecorated tools were made, but that the finely carved and frequently dated Dutch, Austrian, German, and French examples were prized for their artistry and so preserved.

Decorative artistry is not, however, an attribute of the earliest-known English planes. They have generally straight lines and unadorned surfaces. The smaller varieties for smoothing and molding have no handles. These features were carried over into the American plane from the seventeenth century until the plane stock made from wood gave way to one fabricated in metal.

Planes may be categorized in three broad

FIGURE 55.   Planes in miniature. The violin-maker uses tiny tools in his highly refined woodworking. Fashioned from lignum vitae, these are nineteenth-century examples. Lengths, from $\frac{7}{8}$ inch to $2\frac{1}{16}$ inches.

FIGURE 56.   Smooth plane. This one is fitted with a tote, although the usual smooth plane was not. In rosewood, by the firm Bodman and Hussey of Pawtucket, Rhode Island. Length, $10\frac{5}{8}$ inches.

FIGURE 57. (*Opposite above*). Surface decoration of plane. Typical carving at front of the throat and decorative treatment of sides indicate continental European origin.

FIGURE 58. (*Opposite below.*) Plane with fanciful carving. A Germanic style, with carved throat and chip carving surrounding front handle, in a massive plane for jointing edges of heavy boards. Full length, 43¾ inches.

FIGURE 59. (*Above.*) American planes in fruit woods show the simple lines that follow English plane-making traditions. *Above:* Scotia molding plane. Length, 12 inches. *Below:* Ogee molding plane. Length, 9⅝ inches.

groups: bench planes, molding planes, and special types, of which there were a great number. (Planes in the third category — such as plow, dado, fillister, and sun planes — which have more specialized applications in carpentry and joinery, cooopering, or other trades, are discussed in chapters on the trades; they are noted in this chapter only in terms of their construction or materials of manufacture.)

The bench plane was used for surfacing and edge planing of wood held in a vise or temporarily fastened on top of a workbench. American wooden and iron bench planes varied in size according to function; the table below provides representative dimensions.

An extra-long jointer plane (one in the author's collection measures over thirty-seven inches) is identified by Henry Mercer in *Ancient Carpenters' Tools* as a floor plane, used for leveling floors and jointing the edges of floorboards. It would be particularly useful for the latter purpose (Figure 61).

The general class of molding planes includes planes with a wide range of cutting surfaces de-

signed to create decorative wooden shapes for cabinetmaking and for exterior and interior joinery of buildings. The length of the typical wooden molding plane was standardized about 1800 at nine and one-half inches, a reduction from earlier models, which were as much as twelve inches long. They were approximately as wide as, or slightly wider than, the cutting iron they had to accommodate, the selection of iron depending on function and the shape of the planing surface. The simplest types of molding planes were the hollow and round. In American plane-making terminology, the hollow plane shaped a convex surface and the round plane made a concave cut. This is the only instance in which the name given to the plane is the reverse of the type of molding it would strike. The ogee molding plane, for example, formed an ogee shape, although the bottom surface of the tool and its cutting iron were of an opposite pattern, that is, shaped in a "reverse" ogee. Hollows and rounds were made in pairs, usually in fifteen sizes of cutter widths from one-quarter to two inches, in gradations of one-eighth inch; these

BENCH PLANES

NAME*	LENGTH IN INCHES	WIDTH OF IRON IN INCHES	PURPOSE
Smooth	7–9	2–2¼	Final surface planing
Jack	14–16	2⅛–2¼	Rough, initial surface planing
Fore	20–22	2⅜–2⅝	A "short jointer" for edge planing; also, up to 26 in., called a "trying plane"
Jointer	26–30	2½–2¾	Edge planing of boards

* The bench plane names above are consistently given in these sources: *Catalogue and Price List of Bench Planes and Moulding Tools . . . Arrowmammett Works*, 1858; *Hart, Bliven & Mead Manufacturing Co.'s Catalogue and Price List*, 1873; and *Price List of Genuine "D. R. Barton" Planes, Edge Tools, &c*, ca. 1900.

PLANES

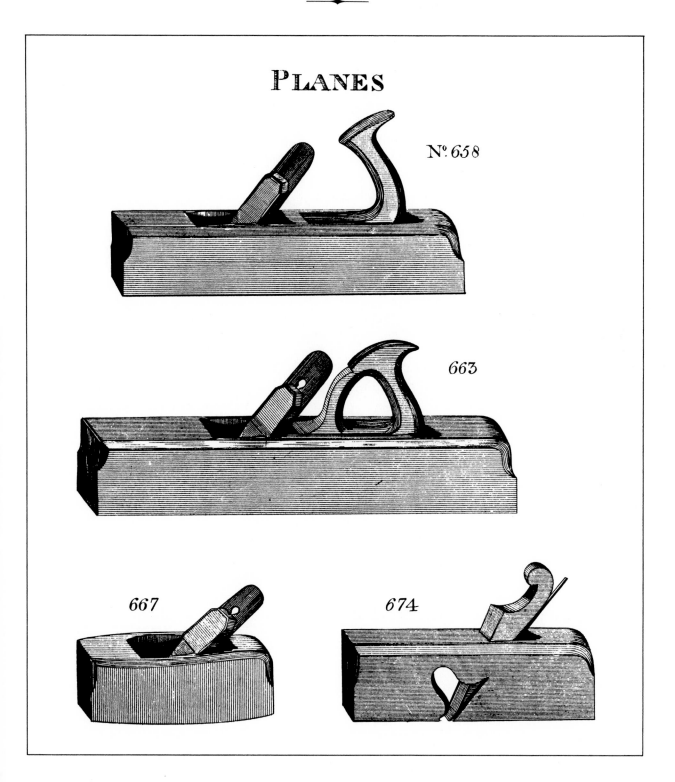

Nº 658

663

667

674

FIGURE 60. Jack plane; fore, or trying, plane; smooth plane; and rabbet plane. From Smith's *Key to the Manufactories of Sheffield*, 1816.

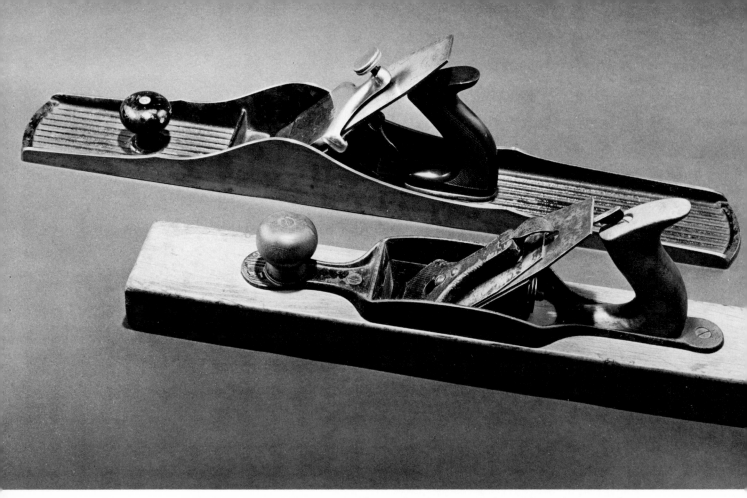

FIGURE 61.   From wood to iron. *Below:* Stanley no. 30 short jointer, incorporating cast-iron fitting screwed to beechwood stock, made from 1870 to 1918. Length, 22 inches. *Above:* Chaplin's cast-iron jointer, patented in 1872 and 1876, made by Tower & Lyon, New York City. The sole is corrugated to reduce friction. Length, 24 inches.

planes worked circles with diameters of from one-half inch to four inches, respectively. The irons were normally set square to the direction of cut, although some of the larger sizes frequently were offered with irons set at an angle, or skewed.

Profile drawings of a representative group of molding planes, showing the configuration of their soles (the reverse of the moldings they would shape), were included in the pattern book published in England by Joseph Smith in 1816. This book, entitled *Explanation or Key, to the Various Manufactories of Sheffield, with Engravings of Each Article*, was published as a salesman's and wholesaler's stock book, probably to show local hardware merchants the breadth of products available. It covers the tools of many woodworking trades, as well as the ex-

tensive lines of Sheffield cutlery. A price list for the work enumerates scores of planes. Among the molding planes offered were beads, quirked beads, fluting, grooving, ogee, reverse ogee, ovolo, reeding, astragal, and many other shapes, and combinations of shapes such as quirk ogee and astragal.

The cornice, or "crown," molding plane was of outsize dimension. It was made from twelve to fourteen inches long and up to six inches wide. Philip Chapin, plane-maker of Baltimore during the 1840s and 1850s, advertised his cornice planes at a price of eighty cents per inch of width. A wide plane of this design was occasionally fitted with two irons, one placed beside but slightly in back of the other and each with its own mouth and throat for the discharge of shavings. Because of the great resistance of the

76

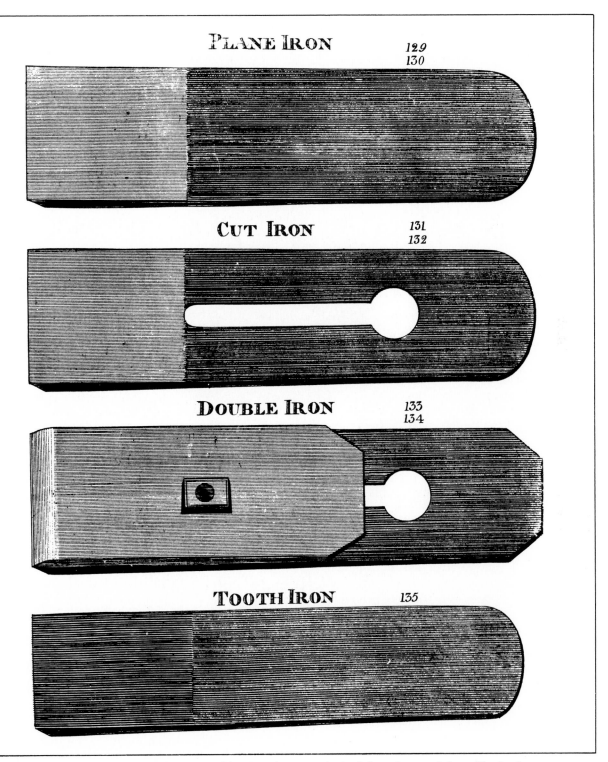

PLANE IRON *129* *130*

CUT IRON *131* *132*

DOUBLE IRON *133* *134*

TOOTH IRON *135*

FIGURE 62.   Single and double plane irons, typical of those imported from England or made in America in the early nineteenth century.

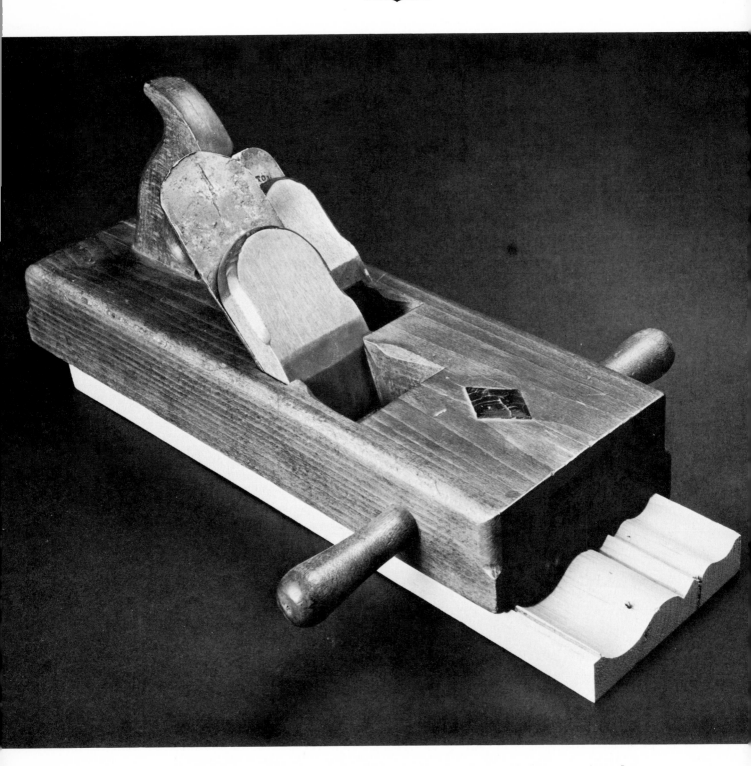

FIGURE 63.   Cornice, or crown-molding, plane. Because of the resistance of wood to its wide cutting irons, two men were needed to operate this 5-inch-wide plane. Length, 14½ inches.

wood to the very wide cutting irons of cornice planes, two supplementary side handles were frequently provided at the front end of the tool. Thus two people could operate the plane, one pushing from the tote at the rear and one pulling the two handles at the front end. The cornice-molding plane was made in a conventional ogee pattern by the plane factories, and early plane-makers designed the working surface of the tool in a variety of patterns to create individualized fancy moldings. The plane was used to form the graceful moldings of exterior cornices below the roof eaves, and the wide interior crown moldings at the junction of wall and ceiling, which were characteristic of the finer homes of the eighteenth and early nineteenth centuries. The cabinetmaker used the tool in a broad ogee pattern to shape the wide framing of mirrors. A soft wood, such as pine, was used for the ogee frame, to which a mahogany veneer was applied.

Until the eighteenth century, when edge-tool makers in England started to manufacture the plane iron as a standard product, it was usually a blacksmith-made part supplied to fit an individual tool. The 1787 *Directory of Sheffield, Including the Manufacturers of the Adjacent Villages* lists twenty edge-tool makers. Of these, several can be identified as makers of plane irons on the basis of planes, both English and American, that are currently in museums and private collections. In addition to name and street address, the *Directory* also gives the "marks," that is, the die-stamped name and/or symbol used by the makers. The marks of P. Law, Newbould, Weldon, John Green, and James Cam (although Cam is listed as a filesmith) are among those in the directory that are frequently found on the irons of early planes of American origin (Figure 8). Such examples provide evidence of the importation of plane irons at an early date for use in locally produced plane stocks, although many were doubtless brought in the tool kits of craftsmen who emigrated from England. Many eighteenth-cen-

tury American planes, however, have irons without any maker's marks. It is probable that a substantial number of these irons were made by blacksmiths living in the immediate vicinity of the plane-makers. The absence of marks at a time when the Sheffield edge-tool makers were regularly identifying their products would support this assumption.

Following the Revolutionary War and well into the nineteenth century, plane irons were regularly imported from England, and particularly from Sheffield, in spite of local manufacture. The reputation of the Sheffield irons was excellent. As with English saws, it was difficult for American makers of plane irons to overcome the prejudice against the locally made item. And, as with saws, the American product was not always of quality comparable to the imported article. A petition of the "tradesmen, mechanics, and others, of the Town of Baltimore," offered to Congress in April 1789, requested increases in duties on the importation of a substantial list of products. The petitioners stated their purpose as "diminishing the rage for foreign, and of encouraging domestic manufactures." Plane irons (as well as "all kinds of edge tools," and carpenters' and joiners' planes) were among the products for which they sought protective duties.[17] In a report on manufactures communicated to the House of Representatives on December 5, 1791, Alexander Hamilton noted the growing production of various kinds of edge tools for the use of mechanics. To foster local industrial development he recommended extension of the duty to ten percent *ad valorem* on all such manufactured products.[18]

During the nineteenth century a market developed for plane irons manufactured by American edge-tool firms. Buck Brothers in Massachusetts was the major producer, along with the Humphreysville Manufacturing Company, Dwights French & Company, and the Baldwin Tool Company, three Connecticut firms, and the Providence Tool Company in Rhode Island. Letters offering testimony to the excellence of

Baldwin Tool Company irons were printed in the company's 1858 "Arrowmammett Works" plane catalog. These irons were prominently advertised as "Manufactured from W. & S. Butchers, Superior Refined Cast-Steel," demonstrating the continuing use of imported Sheffield material. From 1870, the major share of plane manufacturing became concentrated in a relatively small number of firms that maintained in-house manufacture of plane irons. The Stanley Rule & Level Company, the Auburn, and the Sandusky Tool companies, for example, all fitted their planes with irons they produced themselves, as well as offered irons for sale to smaller firms and to hardware stores for the replacement market. In spite of this American production, the English imported plane iron was still being sold in quantity in 1873. A hardware trade journal reported: "Almost all the planes sold in the United States are manufactured or fitted up here, but there is a large importation still of foreign, and especially English, plane irons."[19] The English firms of William Ash, William Butcher, W. Greaves & Sons, Moulson Brothers, and J. Sorby — all of Sheffield — maintained a ready sale of irons, with the Moulson product having perhaps the greatest acceptance.

Edge-tool makers, whether English or American, used two grades of iron for the cutting iron. A blank of relatively soft iron was forged in the required shape. To the top surface of this, on the cutting end, a layer of higher-carbon steel was welded. This steel overlay extended one to two inches up from the cutting edge. After the completed cutting iron was ground and filed to proper shape, either to a straight edge for the flat sole of a bench plane or profiled for an ornamental molding form, it was tempered and dressed with sharpening stones. The top surface of hard steel served to maintain a durable cutting edge as the cutting iron was worn down in subsequent sharpening. The iron was forged in an elongated taper, thicker at the cutting end and thinner at the top. The taper permitted

easier adjustment of the depth of cut. The user could set the iron so that its edge barely cleared the sole of the plane and then lock it in position by striking the top of the wedge with a mallet. To increase the depth of cut, he lightly struck the end of the iron projecting from the top of the stock, thus advancing the cutting edge, and then tightened it by striking the wedge.

In the mid-1700s a "double" iron for bench planes was developed in England (Figure 62). In 1767 Samuel Carruthers, a Philadelphia plane-maker, advertised his planes as "double-iron'd."[20] In constructing the double iron, a standard bench-plane iron was slotted longitudinally and a "cap" iron affixed to the top of the cutting iron by a bolt passing through the slot and into a threaded boss on the cap iron. The addition of the cap, or top, iron served two purposes: it stiffened the cutter, and it diminished vibration, or "chatter," which created an undesirable finely rippled surface rather than a smooth finish. Its firmness also made it effective in planing the end grain of wood or figured woods such as curly maple. In addition, the cap iron lifted and curled the shaving produced by the cutter without raising the grain of the wood. In the 1860s, Leonard Bailey introduced double plane irons made from steel of uniform thickness, in place of the heavier tapered cutters, for use in the planes with cast-iron stocks which he manufactured. These planes did not utilize a conventional wedge for securing the iron in position, and a tapered cutter was neither required nor mechanically suitable for them.

Plane-making (as opposed to the manufacture of plane irons) as a trade probably began in colonial America in the early eighteenth century. There were three general centers of production: southeastern New England, New York City and the Hudson River valley, and Philadelphia. Information about the early plane-makers is increasing steadily as new names are found, makers' places of residence identified, and their working periods discovered. Style, materials, and dimensions characterize certain planes as

FIGURE 64.   Plane irons. From two crown-molding planes made by Jonathan Tower of Rutland, Massachusetts. The irons were marked "J SMITH" by their maker.

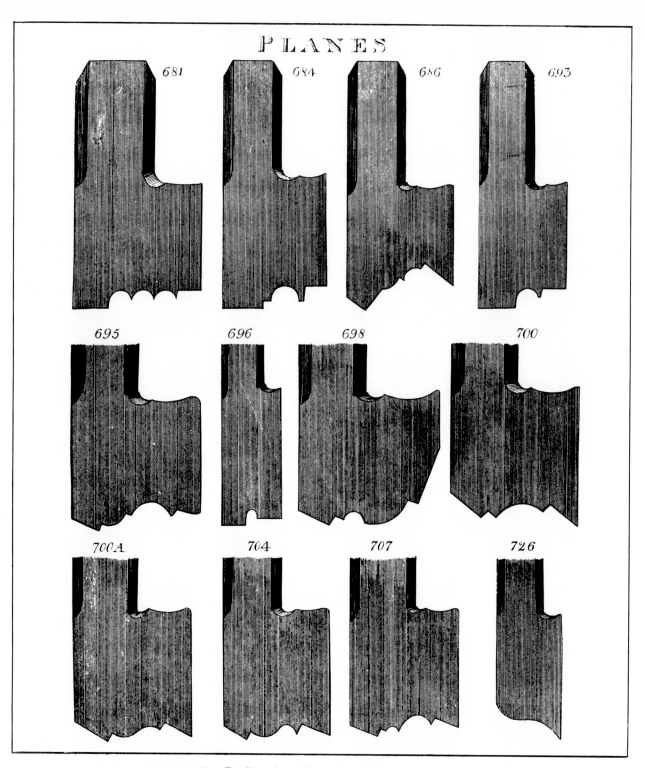

FIGURE 65.   Profiles of molding planes from Smith's *Key*.
681. Triple reed
684. Astragal
686. Bead and cove
693. Side bead
695. Astragal and cove
696. Cock bead
698. Cove and bead
700. Ovolo
700A. Ogee
704. Quirked ogee
707. Quirked ovolo and astragal
726.   Snipe bill

FIGURE 66. Eighteenth-century molding planes, showing makers' marks of Thomas Granford, London; Francis Nicholson and Cesar Chelor, Wrentham, Massachusetts; John Nicholson and David Clark, Cumberland, Massachusetts; A. Smith, Rehoboth; A. Spicer, North Groton, Connecticut; and John Lindenberger, Providence, Rhode Island.

eighteenth-century products. They may afford an opportunity for further identification depending on the presence of an owner's or maker's mark stamped on the nose. Some planes, either homemade or made by a craftsman for sale, have no marks, while others have an owner's or maker's initials. A substantial number of early planes have clearly stamped names of makers who have been identified as working prior to 1800. Finally, there are those planes marked with both the maker's name and town name. This information simplifies the search for the makers' chronological place among American tool-makers and for further historical information about them. By the nineteenth century, plane manufacturers who operated small establishments, as well as the large-scale producers, regularly, although not invariably, identified their tools with both name and location of manufacture.

Plane-making is first recorded in colonial cen-

ters of population, trade, and transportation, at least in the case of Philadelphia and New York. In New England, most of the plane-makers worked in towns clustered in a small area south from Boston and north and east of Providence. These towns were all within thirty miles of the seacoast where the first settlements were established, or were themselves seventeenth-century settlements. The town of Wrentham is at the center of an area that includes the plane-making towns of Boston, Dedham, Middleboro, Rehoboth, Norton, Medway, and Mendon in the Massachusetts Bay Colony, as well as Cumberland and Providence in the Rhode Island colony; Wrentham is no more than twenty-five miles distant from any one of these towns. Of twenty plane-makers known by both name and the town in which they worked and who were in business prior to 1800, fifteen resided in one of the ten towns listed above.

The earliest American plane-maker known

is Francis Nicholson of Wrentham. In wills probated during the colonial period it was not common for the testator to identify himself by occupation. Nicholson, however, in his last will and testament dated April 1, 1752, described himself with pride as "Toolmaker of Wrentham." He died in his seventieth year, and was thus born only some sixty years after the initial settlements in the Bay Colony. His will was recorded on December 14, 1753, in the office of the Court of Probate in Boston where, fortunately, the contemporary transcript still exists. The will and the inventory of his estate filed by his son John as executor provide much information about his property and about Deacon Nicholson as an individual. He wrote: "I Will & bequeath unto my only Son John . . . all my wearing Apparell & Armour, & one third of my Books, & one sixth part of my Cash & Debts due to my Estate, & also all my Tools & timber, except what is before excepted. . . ." The significance of the last words of the bequest lies in the fact that Nicholson owned an indentured black slave, Cesar Chelor, who figured prominently in the will. Chelor, evidently, had learned the trade of plane-making from Francis Nicholson, and a provision of the will reads:

As to my Negro-man Cesar Chelo[r], considering his faithful Service, his tender Care, kind & Christian Carriage, I do set him free to Act for himself in the World, & I do Will & bequeath unto him, his Bedstead, Bed & Beding, his Chest & Cloathing, his Bench & common bench-tools, a Sett of Chizells, one Gouge, one Vise, one Scyth & tackling, & ten Acres of land, to be set off to him at that end of my woodland next to Ebenezer Cowell's, & also one Cow common right in the undivided Land in the Town of Wrentham, to him & his Heirs & Assignes forever, & one third part of my timber.

Along with planes made by Francis Nicholson, several bearing the names of his only son John and of Cesar Chelor have been found. They are all similar in style and show common details. Virtually identical formation and lettering style of the name stamps and the words "LIVING IN WRENTHAM" on many of the tools

of all three makers indicate that the stamps were all made by one individual, either Nicholson himself or a local blacksmith. John Nicholson subsequently moved to Cumberland, ten miles south of Wrentham, for at least two molding planes of his manufacture are stamped "IN CUMBERLAND." Cesar Chelor continued to make planes for many years. Although he died intestate in Wrentham in 1784, the inventory of his estate is in the record books of the Probate Court of Suffolk County. His most valuable possessions were listed as "Sundry Tools" valued at 424 shillings, 4 pence, and some "Old Lumber" at 26 shillings; the latter may well have been the residue of the yellow birch he used in making planes.

Other plane-makers of the early period of the industry in the Wrentham area were Jonathan Ballou, Joseph Fuller, and John Lindenberger, all of Providence, L. Little of Boston, I. Jones of Medway, E. Taft of Mendon, S. Doggett and I. Pike of Dedham, Aaron Smith of Rehoboth, H. Wetherell of Norton, David Clark of Cumberland, and E. Clark of Middleboro.

Jonathan Tower, another early plane-maker whose two-hundred-year-old tools are occasionally found today, was born in Sudbury, Massachusetts, in 1758. His father, Joseph, was a millwright who moved frequently from town to town constructing sawmills and gristmills. Both father and son settled in Rutland, Massachusetts, where Joseph owned and operated a gristmill, while Jonathan's principal occupation was making carpenters' and cabinetmakers' tools. As a young man Jonathan had seriously injured his foot with an adz, perhaps in working on timbers for one of his father's mills. This proved to be an impediment to his service in the Revolutionary War, although he did perform guard duty when prisoners of war were quartered in Rutland following Burgoyne's surrender. It did not, however, impair his craftsmanship, for his planes are superior examples of the plane-making art.

Few New York–area plane-makers of the

eighteenth century have been documented. James Stiles, born in England in 1743, was a carpenter and plane-maker in New York City between 1768 and 1775. In later years he moved up the Hudson River to Kingston and continued working there until his death in 1830. Cornelius Tobey of Hudson, New York, was another plane-maker active in the post-Revolutionary period. It was not until the nineteenth century that members of the trade became numerous in New York City, and even then few of the city firms developed the volume of business attained by upstate New York, New England, and the midwestern factories. The ready availability of imported tools from England and continental Europe in the port of New York may have discouraged development of the trade in the early periods, and later the costs of bringing raw materials into the city undoubtedly contributed to the relative unimportance of the local manufacturing.

Philadelphia, a third center of the trade, had at least nine plane-makers in 1788. A local newspaper of that year provides an account of a stirring "Grand Federal Procession" on the Fourth of July in celebration of the Declaration of Independence and the "establishment of the Constitution or Frame of Government." Military companies, the trades, and professions were represented in the festive parade, which was led by twelve Axe Men. Some distance behind the contingent of four hundred and fifty architects and house carpenters came the plane-makers, led by William Martin "bearing the standard, white field, a smoothing plane on the top; device, a pair of spring dividers, three planes; a brace, a square, and guage [*sic*], followed by eight plane makers."[21] William Brooks, John Butler, Melchior Deeter, Hibsam Martin, and Thomas Napier, a Scotsman recently emigrated from Edinburgh, were Philadelphia plane-makers of the period 1785 to 1800; they may well have marched with William Martin in that Grand Procession.

American planes made during the eighteenth century have characteristics of style that distinguish them from those made after 1800. The length of the earlier molding planes, as previously noted, was not standardized, and they were usually longer than the nine and one-half inches that became standard after 1800. The edges of the wood body of the earlier planes were consistently chamfered with a sloping bevel up to as much as three-eighths of an inch in width. This chamfering ran along the top edges of the stock and extended partway down the toe and the heel. The chamfer was an extra step in manufacturing, but it created a more pleasing design, and the chamfered tool was easier to hold than the later planes with the edges only slightly rounded.

Many of the early plane-makers of the Wrentham area incorporated another small but important feature in their molding planes. They cut a very slight bevel into the front edge of the opening for the wedge, on the top surface of the plane (Figure 67). This seemingly trivial refinement prevented the edge of the slot from splitting if the carpenter or joiner drove the wedge into it too forcefully.

The shape of the wedge is yet another clue in dating (Figure 68). Wedges were taller in the early period. The front edge was cut away in a gradual slope. This gave a more graceful shape to the wedge and also provided more room to strike the underside of the rounded top with a small mallet, to loosen the wedge for adjusting the cutting iron, or to remove the iron for sharpening. In making the wooden wedges for the wide throats of bench planes, the pre-1800 maker regularly rounded the top, and occasionally he chamfered the upper edges. This shaping conformed to that of the plane irons of the period, which were also rounded at the top.

An indication of an eighteenth-century date for a plow plane — used to make grooves — is the use of iron rivets rather than wood screws to fasten the arms of the plane to the "fence," the attachment which served as a guide. Whatever the method of fastening, the wooden parts had a

FIGURE 67. A refinement of early construction. *Above:* The eighteenth-century maker chamfered the wedge slot to prevent splitting. *Below:* Nineteenth-century plane which has split where wedge was driven into its slot.

tendency to dry out and become loose; the arms of the plane would then swivel and fail to maintain firm right angles to the fence. To overcome this problem, the bases of the arms were set into small dadoes cut into the top surface of the fence. The eighteenth-century craftsmen of the Wrentham group used this technique in making plow planes, but it is rarely encountered in tools of the following century (Figure 69).

Finally, the offset tote is also characteristic of the eighteenth-century plane. On many crown-molding planes and raising, or panel, planes, and on some jointers, the handle was set to the right of center on the top of the stock. The reason for this is not known to the author.

The woodlands and forests of East Coast America provided an abundant variety of wood for plane-makers. Planes produced by seventeenth-century English makers are invariably of beech. The Philadelphia artisan Samuel Carruthers also apparently favored beech, for in an

advertisement in the *Pennsylvania Chronicle* in 1768 he informed readers: "N. B. Wanted beach wood, in bolts, also scantling." Carruthers would have used the large bolts of wood to fashion the stocks of large bench planes and the smaller scantlings for small smooth planes, wedges, and other parts. The first New England plane-makers almost invariably used yellow birch; this is the wood found in the planes of the Nicholsons, Chelor, Lindenberger, Fuller, and the Clarks of Middleboro and Cumberland. Yellow birch was common in the area where they worked, but beech was equally plentiful, so they adopted the birch for their tools by choice in the Massachusetts Bay Colony.

By the early 1800s, beech became the standard wood from which plane stocks and wedges were manufactured, and its use persisted in the ordinary grade of plane through the nineteenth and into the early twentieth century, when fac-

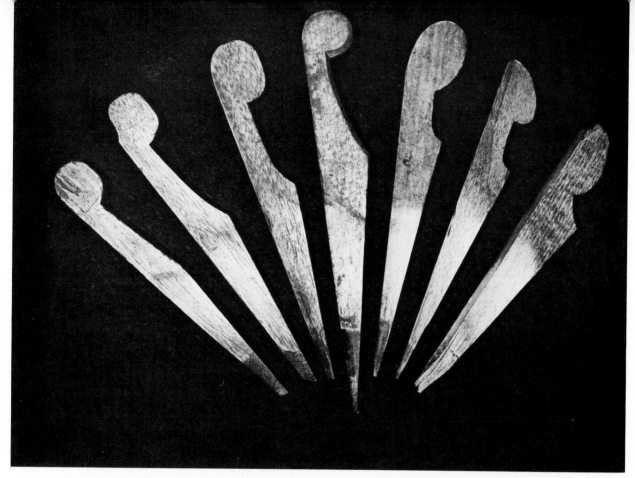

FIGURE 68. Chronology in plane wedges. The first five, at left, are before 1820. After that date the wedge was usually factory made and lost some of its graceful shape.

tory production of wooden planes ceased. Beech offers properties highly suitable for wooden tools. It is hard and fairly heavy, its spring and summer growth produces an even texture, and the lack of open pores and rays makes it easy to work either with, against, or across the grain. But it was not used exclusively, for its physical properties were not suitable for the threaded arms and wooden nuts of adjustable plow planes. For these, boxwood, imported from Turkey, was the principal material used. Boxwood is extremely hard and has almost no visible grain, so it was also the usual choice for "boxing" the surfaces subjected to greatest wear on the soles of wooden molding planes (see Plate 20). In boxing, the sole was slotted longitudinally and hardwood strips of boxwood inserted and held with glue. For extra durability, the wood was sawn so that the grain of the wear strip was diagonal to the wood being planed. The bead plane might be made "single box,"

with one boxwood wear strip inserted, or "double box," with two, or "full box," wherein the entire shape of the molding surface was one large boxwood insert.

For high-quality planes of specialized types, nineteenth-century factories produced planes of wood other than beech. Bench planes were rarely manufactured in other woods, but the plow plane, commonly made of beech with boxwood arms, was also made in full boxwood, rosewood, apple, or ebony. For the most discriminating cabinetmaker, the ebony, boxwood, or rosewood plow plane was fitted with ivory-tipped arms, and even with large ivory adjusting nuts (Plate 6). These top quality tools sold for about five times the price of the least expensive plow plane in beech, yet the beech tools were comparable functionally. Apple or lignum vitae, an extremely hard tropical wood, so dense that it will not float, was used for special planes subject to heavy wear and rough handling.

FIGURE 69. Fastidious workmanship. Base of plow plane arm is set into the fence to position it firmly, a detail that is unusual after 1800.

Ship joiners, the skilled craftsmen who performed the finish woodworking on cabins, railings, and hatchways of the finely fitted sailing vessel, traditionally made a number of planes for their own use, using exotic woods such as lignum vitae. The towns they lived and worked in were ports for importation of the tropical woods of the West Indies and South and Central America; sometimes these woods even came in as ballast or dunnage. Thus they had at hand a ready source of supply and put these dense woods to good use in making smooth, jack, and jointer planes. By tradition they also made the longer bench planes in a "ship" or "razee" pattern, in which the rear of the stock was lower than the front. A graceful curve, following the line of the bottom curvature of the plane handle, effected the transition in the height of the stock. In severe simplicity and grace of line, and in quality and finish of wood, the jointer plane of

the ship joiner is one of the most beautiful of woodworking tools (see Plate 9).

The English revived the use of iron for the manufacture of miter planes and other types of bench planes (Plate 13). Initially the tool was made from wrought-iron plates; the side pieces were fitted to the sole by hand-shaped dovetails characteristic of cabinetmaker's joinery. The cavities of the stock in front and in back of the cutter were generally filled with wood, a use of material and a structural detail that was reminiscent of the Roman plane. Later, the iron body was cast in one piece.

In America, iron was also recognized as a potential material for planes. The first plane-maker of record who used metal for the plane stock was Hazard Knowles of Colchester, Connecticut. Knowles applied for a patent on a cast-iron, rather than wrought-iron, construction. The specifications of his application, on which he was granted a patent on August 24, 1827, proclaimed the "peculiar excellency" of the cast-iron stock: it was more durable than that of the conventional wooden plane; the face of the sole and the slot for the shavings would not wear irregularly; and it could be "afforded at a much cheaper price." It is difficult to measure Knowles's success as a plane-maker. Certainly planes constructed in the design of his patent are extremely rare. A disadvantage of the

FIGURE 70. Introduction of cast iron. Jack plane in the pattern of Hazard Knowles of Connecticut, who was granted first American patent for a cast-iron plane in 1827. Length, 16½ inches.

FIGURE 71. English panel plane. Tool for precise cabinetwork and joinery, and not for shaping raised panels, as was the American panel plane (see figure 92). Length, 17½ inches.

cast-iron tool is that it is subject to fracture when dropped. The attrition by breakage over the one hundred and fifty years since Knowles's patent, and the likelihood that the sales volume, and thus the production, of such a novel tool was small, may well account for the scarcity of his plane today. Knowles remains, however, a significant figure in the history of American plane-making because he carried his concept beyond the patent to a successfully manufac-

tured product. Yet it was to be another thirty to forty years before the production of iron planes assumed significant proportions.

Among many individuals who contributed to the development of the iron plane in the last half of the nineteenth century, Leonard Bailey was outstanding. His first patent, of August 7, 1855, granted while he was a resident of Winchester, Massachusetts, was for an adjustable iron "plane-scraper." The tool was used by cabinet-

FIGURE 72. Bailey cam-action lever cap. Leonard Bailey's device for holding the plane iron in the stock, patented in 1858, supplanted the wooden wedge.

FIGURE 73. Bailey device for adjusting the plane iron. Mechanical regulation of the fineness of cut was made possible by a Bailey invention of 1867. Most planes made today use this hundred-year-old adjusting system.

makers to give previously planed surfaces a smooth and even finish preparatory to applying stain, varnish, or other coating. Bailey invented a system for adjusting the scraper blade which has remained in use in versions of the tool made ever since. One of his most important contributions was the invention of the lever cap for plane irons, which tightened the plane iron by means of a "thumb cam." This device was first illustrated in one of his patents granted on June 22, 1858, but it was only described in a later patent of August 31 for an "Improved method of Securing Plane-irons to the Stocks of Bench Planes."[22] Bailey first applied the cam-action cap to planes with wooden stocks. Within the next few years, however, iron planes with the new device were successfully produced by the Leonard Bailey Company and Bailey, Chany & Co. of Boston.

Bailey's next revolutionary improvement in plane manufacture was the introduction of a mechanism to provide simple and accurate adjustment of the plane iron and regulate its depth of cut. This was achieved by means of a pivoted lever with one end fitted into a slot in the cap iron and the other end actuated by a grooved brass thumb nut rotating on a threaded shaft. Movement of the top iron also moved the cutting iron because they were bolted together. The invention was described in Bailey's patent of August 6, 1867,[23] which was reissued in June 1875.

It was not surprising that the achievements and growing success of this Boston plane-maker came to the attention of the Stanley Rule and Level Company of New Britain, Connecticut, which was then on its way to becoming the

FIGURE 74. Compass, or circular, plane. Made by Leonard Bailey in Boston before his association with the Stanley Rule and Level Company in 1868. Planing of concave and convex surfaces was made possible by regulating the flexible sole plate. Length, 10⅜ inches.

major producer of carpenters' tools in the country. Stanley was to achieve preeminence in tool manufacture by capitalizing on its employees' patents and improvements in tool design and function, by buying out competitive firms with successful product lines, and by large-scale manufacturing and merchandising of quality products. Stanley had its origins in a partnership formed by Augustus Stanley, Gad Stanley, and T. A. Conklin about 1850 to produce boxwood and ivory rules under the name of A. Stanley & Company. Within seven years it had merged four additional companies, including Hall and Knapp, makers of plumbs and levels, to form the Stanley Rule and Level Company.

Leonard Bailey's association with the Stanley firm was formally initiated in an agreement of May 19, 1868, by which Bailey granted to Stanley exclusive rights to manufacture iron and wood bench planes, spokeshaves, and scrapers under seven patents he had been granted from August 7, 1855, to December 24, 1867, and any extensions of these patents. The first licensing agreement originally gave the inventor a five-percent royalty on the prime cost of manufacturing these tools.[24] In later contracts Bailey agreed to manufacture the tools, as a subcontractor, in accordance with a schedule of payments for each type of plane produced. The labor contract for making the planes was terminated by Stanley on June 1, 1875, at Bailey's request, apparently in a climate of mutual ill will and distrust. The license to manufacture on the Bailey patents, however, remained with the Stanley Rule and Level Company.

In the fall of 1875 Leonard Bailey started to manufacture a line of planes and other carpenter's tools in his own right. To these tools he gave the sanguine name "Victor," a trademark for which he obtained Patent Office registration on January 4, 1876. The critical mechanical part of these tools — adjustment of the plane iron — was first made under a patent application filed by Charles H. Hawley, an employee of Bailey, on October 1, 1875. The introduction of the Victor planes brought immediate response in a suit in interference by Justus A. Traut, a Stanley foreman. Traut claimed prior invention and on January 24, 1876, belatedly applied for a patent on the same new mechanical concept as in Hawley's application. Both applications involved using a metal disc with a grooved scroll to create up-and-down movement of the plane iron. The testimony of Traut, Hawley, Bailey, A. Stanley, and other witnesses to the action, taken in the spring and summer of 1876, is replete with contradictory statements, claims and counterclaims, and the implications of industrial spying and piracy on the part of Leonard Bailey. In a decision in December 1876, the Commissioner of Patents found in favor of Traut, concluding that Traut was the original inventor despite his late application. Hawley, nonetheless, in October 1877 was

granted a patent (which he assigned to Bailey) on the basis of a revision of his original October 1875 application.

This action was but the prelude to a further suit in equity brought in 1877 by the Stanley Rule and Level Company against Leonard Bailey, and a motion to enjoin Bailey from producing the Victor planes. The Stanley Company claimed that Bailey was infringing on his own reissue patent of June 22, 1875, on which he had granted Stanley the manufacturing rights. Bailey counterclaimed that the Stanley Company was offering a second line of planes in competition with those they were making under his patents and was not using diligence in selling the Bailey planes, thus reducing his royalties. Bailey admitted under questioning that still another manufacturer, the Bailey Tool Company of Woonsocket, Rhode Island, had been making a "Bailey" plane since 1872 or 1873, although not under the patent in question. The trademark of the Rhode Island firm, registered

FIGURE 75. Stanley no. 9. Cabinetmaker's low-angle miter plane, incorporating Bailey patents, first made in 1870. "For piano makers and workmen in kindred trades." Length, 13¹⁄₁₆ inches.

FIGURE 76.   Gage "self-setting" smooth plane. Made in Vineland, New Jersey, by a competitor of the Stanley firm, which took over its manufacture early in the twentieth century.

by the Patent Office in September 1875, was a battle-axe, and the product line was christened "Defiance." The Justices of the United States Supreme Court "rode the circuit" in that era, and the Honorable Samuel Blatchford, in a Circuit Court decision of June 22, 1878, ruled in favor of the Stanley Company. In 1880 Stanley purchased control of the Bailey Tool Company and its Defiance line, which it carried for only a short time, perhaps only long enough to dispose of the manufactured inventory. The Stanley Company and Bailey reached a *modus vivendi* in spite of their protracted legal battles. Stanley acted as general sales agent for the Victor line between 1880 and 1884, when it purchased outright Leonard Bailey & Company. They last listed the Bailey Victor planes in their catalog of 1888.

The Stanley Rule and Level Company had an immediate and expanding success with the Bailey planes made on the patents secured in the original contract of 1868. These planes were made in iron or in a less expensive "transitional" model with cast-iron upper frame fitted to a wooden stock (Plate 19). The firm advertised in *Iron Age* in 1873 that over sixty thousand Bailey planes had been sold; a year later the figure stood at eighty thousand. In the price list of 1888, sales were recorded at almost one million, and by 1898 a total of over three million had been marketed. Attempts of other manufacturers of iron planes to compete with the Bailey design made by Stanley were rarely successful while the Bailey patents were still in force. The Sandusky Tool Company in Ohio, the Gage Tool Company of Vineland, New Jersey, and the Greenfield Tool Company in Massachusetts, all large-volume producers of planes, as well as scores of smaller firms making tools from wood, wood and iron, or cast iron, disappeared from the scene in the closing quarter of the nineteenth century or shortly thereafter. Leonard Bailey's lever cap with its cam-tightening feature and his basic idea for adjustment of the cutting iron had wholly won the day. These features are standard on the majority of bench planes made even today, more than one hundred years after their introduction.

By 1900 the iron plane had almost entirely supplanted the tool made from wood. The Ohio Tool Company of Columbus, Ohio, and Auburn, New York, one of the latter-day giants in hand tools, still devoted twelve of the thirty-six pages listing planes and plane parts in its *Catalogue No. 23* (about 1910) to "Moulding and Fancy Wood Planes," and an additional nine pages to wood bench planes and special planes. This was a last moment of glory for such hand tools. The Ohio Tool Company, at that time one of few firms marketing wood

planes, was to be liquidated within ten years.

The Bailey patents owned by Stanley had expired by the first quarter of the twentieth century. Sargent and Company and the Winchester Repeating Arms Company, both of New Haven, as well as the Ohio Tool Company, produced extensive lines of iron planes that were similar to the Stanley product. But they came too late on the scene with their copies. Stanley had the established position in the trade, and continued to meet the needs of the American market for iron hand planes; the other manufacturers could not compete and abandoned their hand tool lines. Planes were in a diminishing market, restricted by the introduction of power hand tools and by technological change.

■——— THE DRAWING KNIFE ———■

Examples of the drawing knife — or draw knife, or draw shave — were found among the ruins of Pompeii, yet no medieval example has been found, nor any representation in medieval art (as Goodman notes). It is evident that the tool, for unknown reasons, passed into relative oblivion for many centuries. One of the earliest illustrations of this two-handled implement appears among the tools of the carpenter in Moxon's first essay on carpentry, published in 1679. It was used, according to Moxon, for shaping legs of small stools and ladder rungs. Although he states that it was "seldom used about Home Carpentry," it would certainly have been the obvious tool to shape the one-and-one-half-inch round pins he later describes for joining the mortises and tenons of house timbers.

The tool made and used in America took the form of an iron blade, from about six to twenty inches long, with projecting tangs extending at right angles to the blade. The tangs were in-

serted into hand-shaped or lathe-turned handles, which were fastened by bending over the extending tips of the tangs, by riveting, or by using nuts on threaded tangs. The basil of the standard drawing knife is on the top of the blade. In use, the blade is held almost flat to the surface of the work and drawn toward the body. Increasing the angle at which the tool is held creates a deeper cut.

In American catalogs of the woodworking trades, the drawing knife is usually listed first as a carpenter's tool, but it also appears among the tools of the cooper, coach- and carriage-maker, wagon-maker, wheelwright, and shipwright. It was the usual implement for hand-shaping tool handles for axes, adzes, hammers, and the like. It was also used in making shingles for roof and wall covering of early American houses. The carpenter split pine, oak, or cedar logs with froe and froe club into rough shingles, or shakes; he then tapered these with

FIGURE 77. Hoop-making at Colonial Williamsburg. Cooper's apprentice, seated at a shaving horse, shaping wood for barrel hoops with the drawing knife.

the drawing knife, leaving the butt end, which was exposed to the weather, thicker.

In shipbuilding, the drawing knife in a standard size of ten to fourteen inches long overall was the tool used for shaping round treenails for fastening joints and planking; and a much larger size was used to rough-shape masts and spars prior to finishing with a plane (Plate 8).

The width and shape of the blade identifies the use of a particular knife (Figure 79). The carriage-maker's had a narrow blade, three-quarters to one inch in width. The blade of the carpenter's knife, frequently forged and ground in cross section like a razor blade, was one and

one-quarter inches wide. Other knives were up to one and one-half inches wide, and the heavier wagon-maker's knife as much as one and three-quarters inches wide.

Although the drawing knife may not have been illustrated prior to the late seventeenth century, it appeared somewhat earlier among tools listed in the estates of Plymouth and other colonial craftsmen. The will of William Lotham, a shipbuilder and probably a resident of Saybrook, Connecticut, who died in 1645, included a "draweing knife." (Significantly, Lotham, in addition to all sorts of lumber, pitch, and other nautical construction materials, had

95

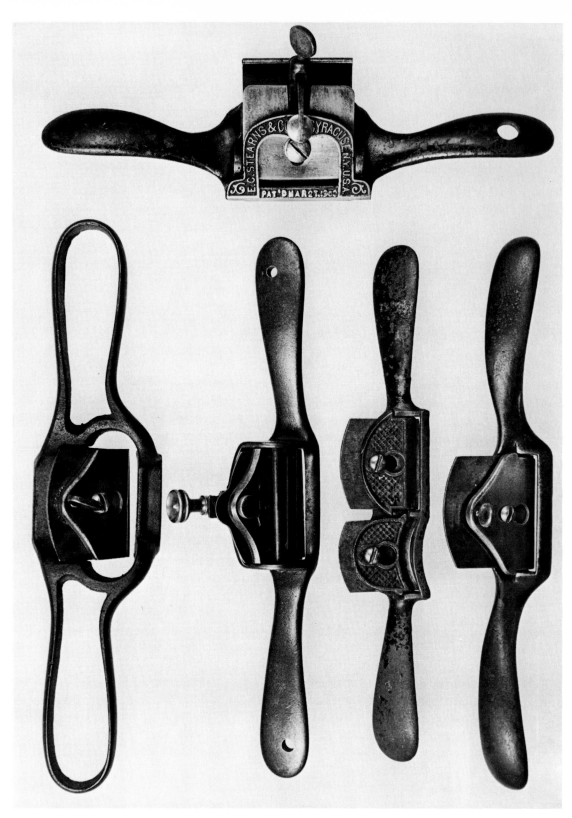

FIGURE 78. Spokeshaves. In iron, as made by Stanley, Leonard Bailey, and other firms. The style at top, with flexible sole to adjust to curved work surfaces, made by Stearns & Co., of Syracuse, New York. Lengths, 9¾ to 10½ inches.

on hand 6000 treenails.[25] Like nail-making at the home forge and anvil, and sap-bucket-making in the lean-to, treenail-making was probably an indoor, winter occupation.) The drawing knife appears seven times in the inventories in *Will [book] &c. No. 2, Sep 12 1716 to Jan 7 1728/9*, in the probate records of Providence. One instance is the will of Major Thomas Fenner, who died in Providence on April 19, 1718. Fenner was a man of varied interests and considerable wealth. His "moveable Esstate" was valued at £433, and included surveying instruments, four oxen, a number of books, a steer, a bull, forty-seven cows, two horses, four mares, and a yearling colt. Among the many tools he possessed were a drawing knife and a howel; listing of the knife in association with the howel, a cooper's tool, suggests barrels were being made by, or more likely for, the major.

Drawing knives were widely manufactured in America by large nineteenth-century edge-tool firms, such as D. R. Barton of Rochester, L. & I. J. White of Buffalo, the Underhill Edge Tool Company of Nashua, New Hampshire, and the James Swan Company of Seymour, Connecticut. One of the earliest-known tool and hardware dealer's catalogs, issued by Wm. H. Carr & Co. of Philadelphia in 1838, lists drawing knives made by N. P. Ames, S. C. Bemis & Co.,

and "various makers." The tool was also commonly made by small, local edge-tool firms throughout the country, by village blacksmiths, and at the farm forge from wrought-iron stock or worn-out files.

The spokeshave, which was suitable for removing material on rounded surfaces and for light chamfering, had characteristics of both the plane and the drawing knife. Like the former, the blade was set into the bottom working surface. Like the drawing knife, it typically had a horizontal, razor-edged blade; other blades were made in either convex or concave shape, for hollowing and rounding applications. Again like the drawing knife, it was used two-handed, but as often as not it was worked away from rather than toward the body. The tool was as small as three to five inches in length for fine pattern-making and model-making work. The largest styles, for coopers, were up to nineteen inches long, with a four-inch iron. The stock of the standard spokeshave was ten inches long, and it held a three-inch blade. Its small working surface was slightly rounded, front to back, in some styles. In others, chiefly those made from cast iron or brass, the sole was flat and held a cutter resembling a miniature plane iron set at a low angle. Because of its versatility for light shaving, the tool was used by wheelwrights, cabinetmakers, joiners, and other artisans.

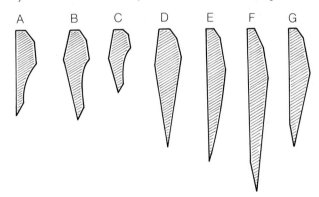

FIGURE 79. Shape of the drawing knife varied according to the task. Cross sections:
A. Carpenter
B. Coach-maker
C. Carriage-maker
D. Wagon-maker
E. Shingle
F. Heading
G. Hoop

IV

CARPENTER AND SAWYER

SHELTER WAS a fundamental need of the early pioneers in America. The colonists, including those of the Virginia plantations and the first Pilgrim contingent at Plymouth, initially made temporary, make-do shelters, which were in many instances merely caves dug into the ground or excavations in a sloping hillside. Within a short time, however, they were constructing dwellings framed with hand-hewn oak timbers. The homes were simple in plan and design, and crude in workmanship; they were the expression of a fundamentally utilitarian approach to solving the problem of shelter with the available handcraft skills, tools, and woodland resources.

There was a plentiful supply of wood in the new land but a serious shortage of both handcraft skills and tools. Of the one hundred passengers who came on the *Mayflower*, forty-one were men. The only one of them who was identified with a woodworking trade was a cooper, John Alden. He was a last-minute recruit, hired when the ship, en route from the port of Delftshaven, put in to Southampton for provisioning. The dearth of men skilled in the building trades accounts in part for the primi-

tive nature of the first dwellings of the early settlers. By 1624 sawyers, as well as a ship carpenter, had joined the Plymouth Plantation, but a shortage of craftsmen was a persistent problem. In 1633 John Winthrop, Jr., son of the governor of the Massachusetts Bay Company, was offering free passage to carpenters from London to induce them to emigrate.

The shortage of tools was also a problem. In 1631 Governor Winthrop, writing to his son, then in London, of the "mainy troubles and adversityes" of the struggling settlement, asked him to bring to the colony "Chalk and Chalke-line, and a paire or 2 or more of large steele Compasses."[1] These tools would be used to mark and measure logs in hewing timbers, not only for new homes, but to replace those which, as Winthrop wrote, had been lost in fires. Fires were a constant problem, because chimneys of the houses were framed of wood and merely lined with dried mud or clay. A book written in 1634 to "benefit the future Voyager" to New England advised the emigrant to take "All manner of Iron-wares, as all manner of nailes for houses . . . all manner of tooles for Workemen . . . with Axes both broad and pitching-axes. All manner of Augers, piercing bits, Whip-saws, Two-handed saws, Froes, both for the riving of Pailes and Laths, rings for Beetles heads, and Iron-wedges."[2] The author acknowledged that these tools were being pro-

FIGURE 80.   Hand saws. Collection of crosscut, tenon, rip, and dovetail saws in the cabinetmaker's shop, Colonial Williamsburg, Williamsburg, Virginia.

duced by blacksmiths in the settlements, but suggested that it would be advisable, nonetheless, to transport them. This was a wise admonition, for tools remained in short supply throughout the colonial period. A "wright," or house carpenter, as a contemporary English document indicates, would have used the whipsaw (a two-man, narrow-blade pit saw), axe, hatchet, adz, hammer, pincers, iron "dog" (a heavy iron staple for holding timbers in hewing), spokeshave, and auger.

The documents of Governor Winthrop provide evidence not only of the importing of tools from England, but of their manufacture in the colonies. The governor's son, in the account book kept for 1630/31 in London, recorded purchasing iron and a pair of smith's bellows for a forge. On July 28, 1631, a pair of smith's bellows was put aboard the ship *Lion*, shortly to sail for New England. The ship *Jonas* carried ten hogsheads and chests of tools for blacksmiths in March two years later.[3] The blacksmith was a key craftsman, for in addition to forging items of domestic hardware such as hinges, latches, and nails, he produced tools for the woodworking trades.

Although the pioneers were greatly dependent for many decades on importation for tools, seeds, animals, and other merchandise, the records show continuing evidence of their striving toward self-sufficiency. The London Port Book for 1634 carries an entry for the shipment of fourteen firkins of nails and four bundles of saws aboard the ship *Elizabeth and Dorcas*. A year later nails could be produced in the Massachusetts plantation: a dozen nailing hammers (the blacksmith's hammer for forging nails from iron rod), and four nailing stakes (small anvils on which nails were formed) were shipped to Boston.[4] Nails continued to be a scarce commodity and in great demand, and the settlers assiduously culled the wreckage of burned buildings to recover them for re-use.

At the time of the colonization of America, carpentry and joinery were separate occupa-

tions. The carpenter framed houses, built the stairs, laid the floors, and in general performed the rough woodworking. He worked to less exacting standards than the joiner, and used heavier and fewer tools. The joiner, in simplest terms, made finely fitted joints. He worked with a variety of planes, chisels, saws, gauges, mitering tools, and patterns to prepare paneling, moldings, door and window framing, stair rails, chimneypieces (the wooden mantels and architectural trim surrounding fireplaces), cupboards, cabinets, and similar refined work to relatively exacting standards. Writing in 1679, Joseph Moxon suggests that the craft of carpentry developed before that of joinery, for "Necessity (the Mother of Invention) did doubtless compel our Forefathers in the beginning to use the conveniency of the first, rather than the extravagancy of the last."

Moxon is extremely valuable for his descriptions of the tools and techniques in use at the time America was being settled. In the portions of his book devoted to house carpentry, Moxon identifies in illustration and text about twenty-five tools, measuring instruments, and mechanical devices. Of these, twelve are pictured in his plate 8 (reproduced, Figure 81). Although he does not illustrate them, Moxon refers to additional tools used by the house carpenter: the ten-foot rule, two-foot rule, auger, compass, piercer and bit, hatchet, plane, and saws.

A second major source for our knowledge of carpenter's tools about this time was published in Paris between 1751 and 1772: Denis Diderot's monumental *Encyclopédie*, a classified dictionary of sciences, arts, and trades. A comparison of the carpenter's hand tools he illustrates with those in Moxon's *Mechanick Exercises* reveals few changes in the century between their writings. The engraved plates of Diderot are considerably more refined in execution than Moxon's. There are over forty detailed illustrations of different carpentry operations in forming a variety of timber joints, framing a house, and constructing a wooden bridge or a sawmill. Yet few of the hand tools illustrated

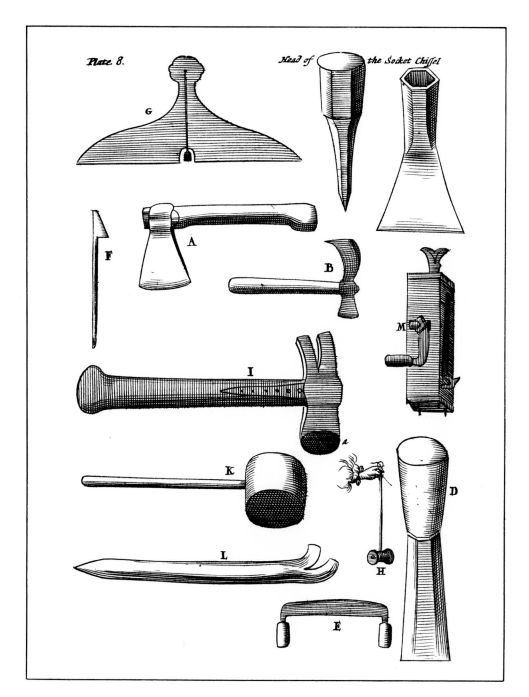

FIGURE 81. Tools of the house carpenter. As illustrated by Joseph Moxon in *Mechanick Exercises*, 1679.

A. Axe
B. Adz
Top Right: Socket chisel
D. Ripping chisel
E. Drawing knife
F. Hook pin, or draw bore pin

G. Level
H. Plumb line
I. Hammer
K. Commander
L. Crow
M. House jack

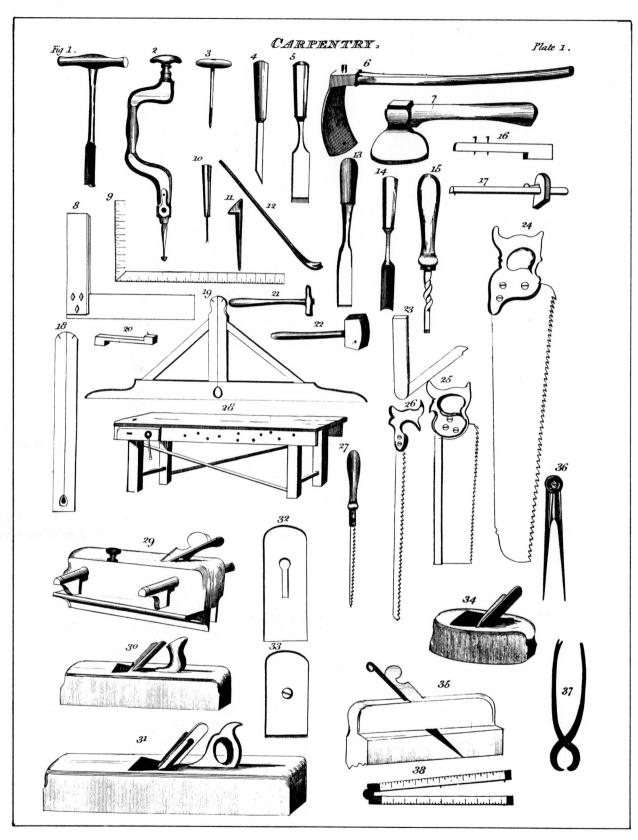

FIGURE 82.   Carpenters' and joiners' tools. From Thomas Martin's *Circle of the Mechanical Arts*, 1818.

PLATE 12. Tool chest of a joiner or cabinetmaker about 1800. Many of the planes were made by an unknown "J. B." from yellow birch.

PLATE 13.   Craftsmanship of Scotland. Low-angle miter plane by Stewart Spiers, of Ayr, ca. 1850. Sides of the metal stock are joined with the sole by dovetails. Rosewood fills the body and gunmetal cap secures the iron. Length, 12 inches.

PLATE 14.   Dado plane, in a styling which is almost architectural, made by Joseph Fuller of Providence, Rhode Island, in yellow birch. Length, 10 inches.

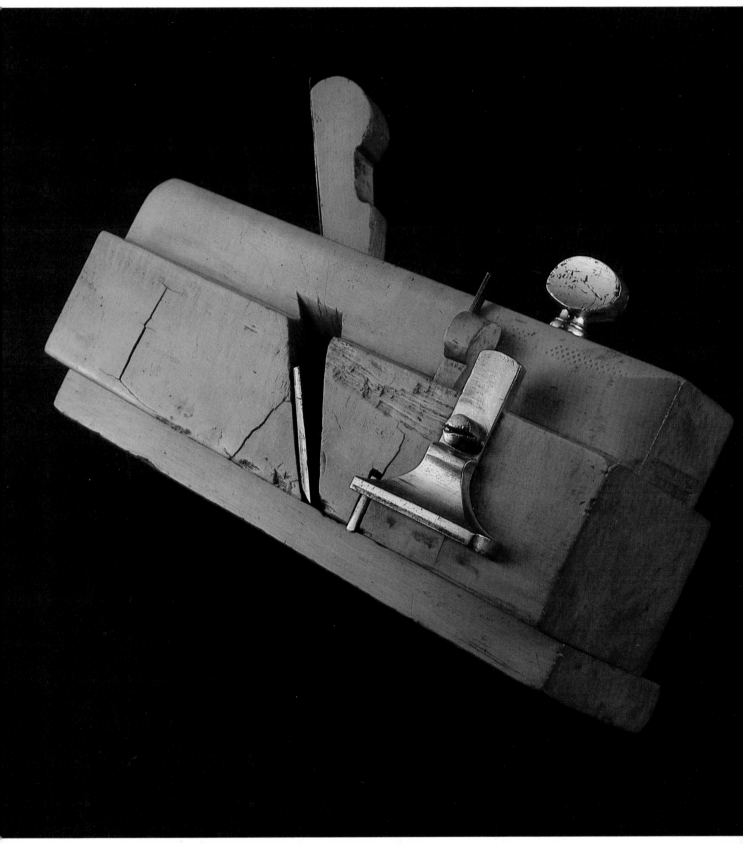

PLATE 15. J. & L. Denison of Saybrook, a small Connecticut company, produced this boxwood moving fillister plane. Length, 9½ inches.

PLATE 16.   Plow plane, by the firm of A. G. Moore, New York City, 1853–1861.
Displays European principle of design in an otherwise American vernacular. The
arms, affixed to the stock, slide through fittings on the fence. (In the American plow
plane, almost invariably arms were attached to the fence.) Beech wood. Length, 11½
inches.

PLATE 17. Adjustable slitting gauge. To cut a thin panel the cabinetmaker used this tool with its knife-edged cutter. Length, 18½ inches.

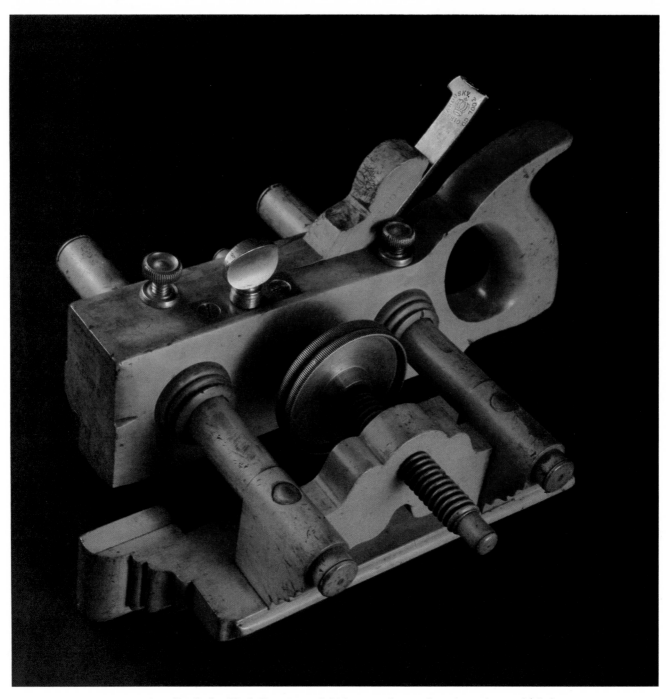

PLATE 18.   Sandusky Tool Company of Ohio manufactured a plow plane which is engineered. The fence is moved by rotation of the brass wheel acting on right- and left-hand threaded shaft. Boxwood. Length, 11⅜ inches.

PLATE 19.   A craftsman produced his own laminated and inlaid stock to fit the metal parts of a Stanley plane made between 1870 and 1885. Length, 15½ inches.

PLATE 20. Boxing. Chiefly bead-molding planes; several illustrate various techniques of boxing, the insertion of boxwood strips at points of greatest wear on the soles.

were not pictured by Moxon. Diderot includes the square, bevel, saw set, chalk-line reel, a scribing tool, and the twibil (a form of double-ended chisel for squaring tenons and paring mortises); all but the twibil, which was a Continental rather than English implement, were included in Moxon's essay on joinery.

Moxon gives a simplified, step-by-step description of framing a house. His plan is for a dwelling twenty by fifty feet. There is a detailed account of how the master carpenter first prepares a scale drawing of the ground plot and floor plans. The early houses of the plantations exhibited applications of the tools and construction techniques essentially similar to those described by Moxon. This is attributable to the influence of a building tradition rather than the use of *Mechanick Exercises* as a builder's handbook. The major early American plantations of New England, New York, and along the Atlantic coast to the south were chiefly outposts of English and Dutch seventeenth-century civilization, and, as might be expected, the technology of early house carpentry and construction clearly testifies to its origins in dwellings of England and Holland.

The typical plan of the earliest seventeenth-century dwellings consisted of one room with a large fireplace at one end, the entire building measuring some sixteen by twenty feet. The frames of some of the first Plymouth houses were not supported by vertical studs between main corner posts; instead, vertical planks were used to create the exterior walls and at the same time provide structural support. Away from the Plymouth area, however, traditional post-and-beam framing with wall studs was the rule. The one-room home was frequently expanded by attaching at one end a duplicate of the original building, with hearths opening from a central chimney into each of the two rooms, or with a chimney at either end of the house. Often a ladder or narrow stairway in the small entrance hallway gave access to the "chamber," the single room on the second floor. The roof was steeply pitched at fifty or fifty-five degrees; the

earliest roofs were covered with thatch, and later ones with wooden shingles. By midcentury, two-room, four-room two-story, and larger houses were being built. Adding a lean-to at the rear, from which evolved the "salt-box" profile, was an obvious and frequently adopted method of increasing living and storage space.

Household estate inventories, taken at the deaths of settlers in the early days of immigration, provide invaluable information and testify to the austerity of their dwellings. Nine inventories of members of Plymouth Plantation who died in 1633, only thirteen years after the arrival of the *Mayflower*, show how limited were their home furnishings, apparel, and household equipment. The lists have pertinence because they reveal how far house construction had advanced from the rude shelters of the first years. From the list and locations given for the possessions of a house carpenter named William Wright, it is possible to reconstruct the plan of his home. It consisted of two rooms on the ground floor, a "first room" and a bedchamber. There was a "loft," probably reached by a ladder, over both of these rooms. On one side of the loft there was an old bedstead, a bag of feathers, and some barrels, while the other side housed a sizable collection of his carpenter's implements. Wright's tools (like those listed in the estate inventories of Francis Eaton and John Thorp, both also carpenters) included the typical assortment necessary for house building: the adz, felling axe for chopping down trees, broad axe for shaping timbers, hatchet, chisel, gouge, plane, auger, piercer (bit stock) and bits, hand saw, two-man pit saw, hammer, chalk line, drawing knife, and file.[5]

Inventories from Hartford and Windsor, Connecticut, of the period 1648–1649, only twelve years after the establishment of these plantations, list belongings located in various rooms and show that some houses had two full stories. These homes were constructed with cellars, chambers or loft rooms on a second story, and "garrits" under the roof peaks.

A letter from Samuel Symonds to his brother-

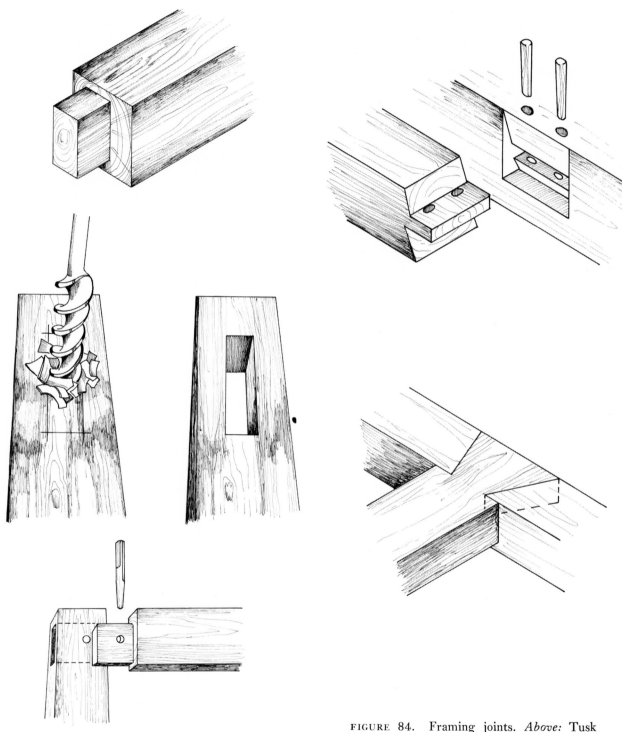

FIGURE 83.   Mortise and tenon. Basic joint in timber framing of the house.

FIGURE 84.   Framing joints. *Above:* Tusk and tenon, joining summer and chimney girt. *Below:* Plain dovetail joint, an alternative method.

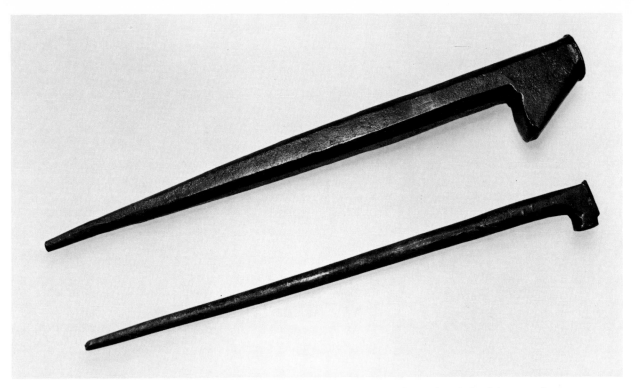

FIGURE 85. Draw bore pins. The smaller, hand-forged iron pin of the joiner, and larger pin of the housewright, for drawing together mortise-and-tenon joints. Lengths, 10 to 10⅞ inches.

in-law John Winthrop, Jr., of Ipswich, Massachusetts, written about February of 1638, affords a most revealing description of the construction of what would have been an expensive two-story home. Symonds wrote:

I am indiferent whether it be 30 foote or 35 foote longe 16 or 18 foote broade. I would have wood chimnyes at each end, the frames of the chimnyes to be stronger then ordinary to beare good heavy load of clay for security against fire. you may let the chimnyes be all the breadth of the howse, if you thinke good, the 2 lower dores to be in the middle of the howse one opposite to the other. be sure that all the dorewaies in every place be soe high that any man may goe upright under. the stairs I thinke had best be placed close by the dore. it makes noe great matter though there be noe particion upon the first flore if there be, make one biger then the other. for windowes let them not be over large in any roome, and as few as conveniently may be. let all have current shutting draw-windowes [casements], haveing respect both to present and future use. I thinke to make it a girt howse will make it more chargeable then neede. however, the side bearers for the second story being to be loaden with corne etc. must not be pinned on but rather eyther lett in to the studds, or borne up with false studds and soe tenented in at the ends; I leave it to you, and the Carpenters.

In this story over the first, I would have a particion, whether in the middest or over the particion under I leave it; In the garrett noe particion, but let there be one or two lucome [luthern, or gable] windowes, if two, both on one side. I desire to have the sparrs [rafters] reach downe pritty deep at the eves to preserve the walls the better from the wether. I would have it Sellered all over, and soe the frame of the howse accordengly from the bottom. I would have the howse strong in timber though plaine and well brased. I would have it covered with very good oake-hart inch board *for the present*, to be tacked on onely for the present as you tould me; let the frame begin from the bottom of the Seller, and soe in the ordinary way upright for I can hereafter (to save the timber within grounde) run up a thin brickworke without. I think it best to have the walls without to be all clapboarded besides the clay walls.[6]

How did the house carpenter construct the early dwellings, and how did he use the tools identified with his trade? His first step after preparing a plan was to fell trees and hew timbers for the sills. Trees, almost invariably oak, were chopped down with the felling or "falling" axe and squared with the broad axe to make the timber frame. In squaring logs the

105

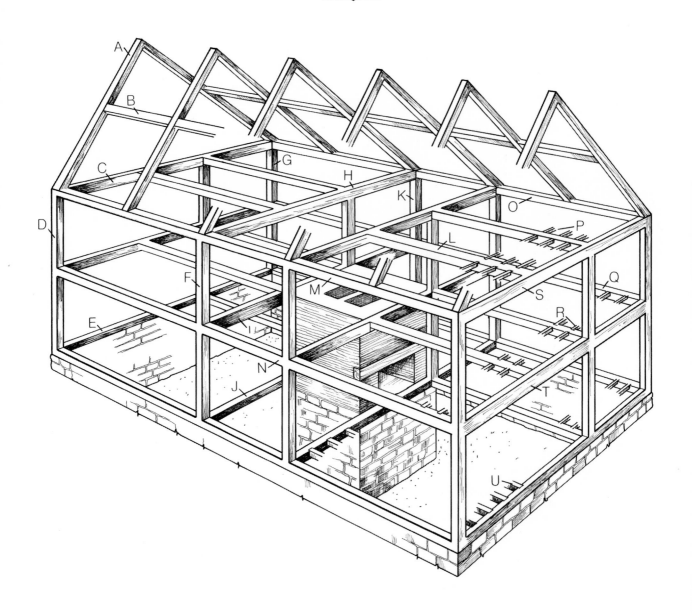

FIGURE 86.   Timber framing of the two-story colonial dwelling.

A. Common rafter
B. Purlin
C. Second end girt
D. Front corner post
E. Sill
F. Front chimney post
G. Rear corner post
H. Second chimney girt
 I. First chimney grit
J. Cellar girt
K. Rear chimney post

L. Second summer
M. Front plate
N. Front girt
O. Rear plate
P. Attic floor joist
Q. Rear girt
R. Second floor joist
S. Second end girt
T. First end girt
U. First floor joists

FIGURE 87. Pit saw. New England sawyers made planks and boards with this huge
framed saw. Length, 7 feet, 4 inches; saw
blade, 5 feet, 3 inches.

carpenter followed a method that had been practiced for centuries. He first marked on both
cross-sawn ends of the log the square or rec-
tangle indicating the dimensions intended for
the finished timber. He then stretched a chalk
line from each corner point of the square or rec-
tangle to the corresponding point on the other
end of the log and snapped the cord to mark a
straight line along the bark. With the felling
axe he cut vertical incisions in the rounded sides
of the log, and with the broad axe cut away the
sectors on all four sides by hewing vertically to
the chalk line markings. The use of chalk to
"whiten a Line, by rubbing the Chalk pretty
hard upon it," and snapping the line to mark the
wood with a perfectly straight line as a guide
for sawing, or equally well, for hewing, was explained fully by Moxon.

The use of axe-hewn framing timbers persisted long after the introduction of water-
powered sawmills in America. Tradition and
habit undoubtedly played some part in the carpenter's spurning the mill-sawn timbers; in any
case, he was often less conservative about
smaller members and would use power-sawn
lumber — for studs, for example — in the same
house with hand-hewn posts and sills.

Sills measured from eight to ten inches in
width and six to eight inches in depth. They
rested on wood pilings, or on stone or brick
foundation set below grade, and were joined at
the corners of the house by heavy mortise-and-
tenon joints. The carpenter used at least nine
tools to fashion this basic joint. With a scribe
and square he laid out at the end of one timber
the lines to which the mortise was to be formed.
With an auger whose diameter was equivalent
to or slightly smaller than the width of the
mortise, he bored holes through the timber and
removed excess wood. Next, he used chisels to
shape the sides and corners of the rectangular
mortise. With the auger he then bored across
the mortise, making a hole about one and one-
quarter inch in diameter for the oak peg he
would later insert. To form the projecting tenon
and its shoulders on the connecting sill timber,

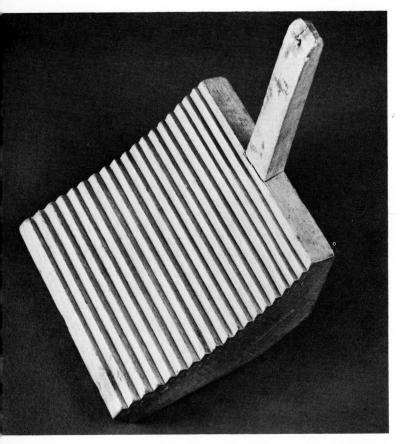

FIGURE 88. Clapboard gauge. This tool scores a line on the clapboard to guide placement of the next overlapping board. Width, 5³⁄₁₆ inches.

he sawed just outside his construction lines. The surfaces of the tenon were smoothed with a chisel until it made a snug fit in the mortise, and the two members were driven together with the heavy, long-handled beetle, or commander. (The head of the beetle was normally bound with iron rings to prevent splitting in this heavy work.) Once the tenon was in position, the outline of the hole bored in the mortise was scribed on it. The timbers were then separated, and the auger was used to bore a hole through the tenon; this hole was not centered precisely on the scribed mark, but was bored slightly closer to the shoulder of the tenon ("about the thickness of a shilling," Moxon suggested). The timbers were once again joined, and an iron draw-bore, or hook pin, was driven through the holes of mortise and tenon, drawing the tenon firmly into the mortise, making a tight fit (Figure 85). Finally, the carpenter knocked out the hook pin and drove in a permanent wooden pin, which he had shaped with his drawing knife to a slight taper.

The "through" mortise-and-tenon joint, in which the aperture for the tenon was cut through the full width of the timber, was a fundamental structural technique in carpentry and joinery. The carpenter used many variations in forming joints, such as the stopped mortise-and-tenon, in which the tenon fitted a mortise cut only partially through the timber; plain dovetail; shouldered dovetail; half dovetail; tusk-and-tenon; and double tusk (Figure 84). The hand saw and chisels were the tools necessary for making these joints.

The framing of the house above the sill consisted of corner and chimney posts, horizontal girts at the second-floor level, plates at the front and back of the roof line, the transverse summer beam, floor joists, and vertical studs. As the timbers of the house were completed and fitted, the carpenter scribed duplicate Roman numerals on adjoining pieces to identify the timbers forming each joint. The house timbers were raised in sections, or "bents," which were assembled on the ground insofar as possible, just as in barn raising, and were consecutively numbered. The process was a community affair, and the participants used ropes, winches, and long wooden pike poles with iron points to elevate each bent to its vertical position. To attain "plumb" construction, with the framing in true perpendicular, the builder tested the vertical alignment with his plumb bob, a lead or other weight suspended by a string from the top of the frame (Plate 3).

A number of variations can be found in the method of framing the roof. For the "common" rafter design (pairs of roof timbers of equal length), the rafters were spaced three to four feet apart, joined by a mortise-and-tenon joint at the peak, and notched into the plate at the eaves line; no ridgepole was used in this design. In one frequent variation, rafters were spaced eight or nine feet apart; horizontal purlins were set into notches in the upper edge of the rafters, the topmost purlin forming the ridgepole. All these posts, beams, and rafters had been prepared with axe and broad axe.

The first carpenters in America not only hewed the framing timbers for houses, but sawed the planks for flooring and roofing and the thinner boards for wall panels and doors. Soon, however, men with the specialized trade of sawyer arrived, and the carpenters could take advantage of the sawn boards and planks which they provided.

Two sawyers normally worked as a team. To raise a large timber for sawing, they supported it on two high trestles, propped it up on one trestle, or set it directly over a pit dug in the ground. In the last method, one man stood on the timber and the second man stood in the pit. The "pit man" pulled the saw down — the cutting stroke — and the "top man" pulled it back up. The pit saw took two forms: open and framed. In the former, the blade of the saw, which was from five to eight feet long, was fitted with an upper handle called the tiller and a lower handle called the box. In the framed pit

saw, the blade was held in tension in a rectangular wooden frame with handles at each end. Ash was a favored hardwood for making the saw frames.

A sawyer must have been working for Governor Theophilus Eaton of Connecticut when the governor died in 1657. In his estate were listed, in addition to two whipsaws (pit saws) and a block saw, "a planke upon Tressills, & some sawed boards."[7]

From an early date the colonists in New England used horizontal oak clapboard siding over the posts and studs of walls. They built up the walls between the posts with "cob," a mixture of clay and chopped straw. Clapboard siding was not used commonly in England, but it is supposed that its use became general in America because the mud and straw filling between timbers, if left exposed as in most English houses of the period, would not withstand the driving rains and severe winter storms common to the Atlantic coast region. However, clapboarding did, in fact, reflect a construction method with which the first Pilgrims and other settlers from England were quite familiar. The majority of the inhabitants of New England in 1640 had their roots in Suffolk, Essex, Norfolk, and other southeastern counties of England, where seventeenth-century houses were built of post-and-beam design, and in many instances closed in with horizontal clapboards of oak or elm, with roofs of thatch. Moreover, many of the Pilgrims had lived for a period of twelve years and more in Holland, in Amsterdam, Antwerp, and Leyden, cities where the timbered house with wooden siding was not uncommon.

Clapboards were at first riven with the froe, a splitting tool, following the pattern of rays extending outward from the center of the log. They were therefore "feathered" — that is, narrow at the top and wider at the bottom, the edge that was to be exposed to the weather. By the eighteenth century, sawn clapboards of white pine or cedar had supplanted the riven clapboard of oak. The carpenter used a gauge to

FIGURE 89. Shaping a gutter. The rounded sole of the plane smooths the trough of a wooden rain gutter. Length, 15⅝ inches.

score or mark a line on each clapboard so that he could maintain regular spacing. This gauge might be a crude one he had made for the job himself or, much later, a manufactured tool. The length of clapboards was determined by the spacing of the vertical studs to which they were nailed; no sheathing boards were used underneath. Roofing shingles, on the other hand, were nailed over wide, roughly finished and unplaned boards which covered the roof rafters. Shingles were split with the froe in lengths from fourteen inches up to three feet (the latter a size unknown today), and in six- to eight-inch widths. A typical price in the mid-seventeenth century, set by the New Haven Court in 1641, was nine pence per hundred for two-foot-long shingles with planed edges. Oak was the usual wood for shingles at that time.

The interior flooring of oak or pine, at least one inch thick, was frequently laid over a subfloor of wide boards nailed across the floor joists. The carpenter often used finish floorboards as wide as eighteen or twenty inches. The edges were planed smooth with the long jointer plane and surface planed, but tongue-and-groove edges were not used.

The *Records of the Governor and Company of the Massachusetts Bay in New England* offer a striking account of efforts to control prices and wages within the first twenty years of the settlement.[8] Inflation caused by a scarcity of labor, and alternate deflation as a result of a lack of money in circulation, brought tribulation to the governing authorities. They attempted to effect controls by first establishing, then abolishing, and then again renewing specific rates of pay in a series of acts which foreshadowed the efforts of our contemporary governments to maintain economic stability.

In August 1630, at the initial session of the Bay Company's Court of Assistants, workingmen's pay was an immediate item of business. The court ordered that carpenters, joiners, bricklayers, sawyers, and thatchers were not to take more than two shillings a day, nor was any man to pay more, under pain of a ten-shilling fine to both taker and giver. The pay of sawyers was limited to "4s 6d the hundred for boards, att 6 scoore to the hundred, if they have their wood felled & squared for them, & not above 5s 6d if they fell & square their wood themselves." In September of the same year, the court regu-

FIGURE 90. Pump log tools. Tapered reamer, for inside female taper of one end of a log; "sheep's-head," or outside tapering tool for opposite end; and bit section of an 11-foot-long auger for boring water-pipe logs. Length of handles, 15 to 16 inches.

FIGURE 91. Pump plane. Wooden backyard-pump cylinder is formed by two matching sections shaped by the semicircular cutting iron. Length, 14⅞ inches.

lated the pay of master carpenters, joiners, and other craftsmen, at not more than 16d per day "if they have meate and drinke." By March of 1631 the court had removed restraints on pay, ordering that the wages of carpenters, joiners, and other artificers and workmen "shall nowe be lefte free & att libertie as men shall reasonably agree." But in October two years later the court reimposed a wage scale of two shillings a day for carpenters, sawyers, clapboard rivers, joiners, wheelwrights, and other artisans "finding themselves dyett," and not more than fourteen pence if food was provided. As usual, fines for noncompliance were to be imposed. In September 1635 the court in desperation once again abandoned wage controls, and in 1636 local option was instituted, "the freemen of every town to agree amongst themselves about prices and rates of all workmen." Finally, in June 1641 the General Court sitting in Boston took cognizance of the level of consumer prices in setting wages. Commenting on the scarcity of hard money and the great decline in prices of corn, cattle, and other commodities, it declared that workmen should be content to reduce their wages "according to the fall in commodities wherein their labors are bestowed," or accept payment in kind.

The control of wages was a general practice through the early years of the settlements. The General Court of the Connecticut plantations, sitting in Hartford in June 1641, expressed concern about excessive rates, established maximum scales for carpenters, wheelwrights, joiners, and others, and hours of labor as well. Artisans were ordered to work eleven hours daily in the summer and nine hours in the winter, "besides that which is spent in eateing or sleeping."[9]

The carpenter's tool chest of the first half of the nineteenth century contained essentially the same tools it had held during the preceding two hundred years. Inventions had improved the auger. The double iron for planes, a new type of level, and bit braces of iron and wood with chucks to hold any assortment of standardized bits were other improvements now in general use.

A carpenter in the early 1800s might also own several tools developed for special applications. For example, there was a plane made for smoothing the troughs of wooden rain gutters after they had been roughly channeled with a gutter adz (Figure 89). From the time of the first colonial aqueducts and water systems until cast-iron, wrought-iron, lead, and other metal piping came into general use in the nineteenth century, water in town and on the farm flowed through wooden water pipes. Aaron Burr had formed the Manhattan Company in 1798 to install such a system in New York City. As recently as 1926, the small New Hampshire town of New London installed a complete nine-mile water system consisting of staved wooden piping. For the usual nineteenth-century water pipe, special long augers and reaming tools were used to bore holes in ten-foot sections of pipe fashioned from tree trunks. Pumps for outdoor wells were also made of wood before the cast-iron pump became common. The hole in the pump shaft was bored with a pump log auger or shaped with a special pump plane (Figures 90, 91). This plane cut a half-round hollow; two sections thus shaped were then fitted together to form the pump shaft.

Another specialized carpenter's tool was the stair saw. The tool, with its six- to eight-inch blade, was used to saw shallow recesses into the wall string and outer string board into which the ends of the stair treads and risers were fitted (Figure 32).

Developments of the 1800s were to effect great changes in the carpenter's trade, in the types of tools he used, and particularly in the level of skill required of him. Mill-sawn lumber, nail-making machinery, and special-purpose woodworking machines made building construction faster and less expensive; less skillful craftsmanship was needed, and certain tools were no longer required.

Waterpower had been used for sawing timber on the European continent from the early Middle Ages, but the trade of the sawyer persisted for centuries after the introduction of sawmills. In America, the sawyers continued to work in competition with the sawmills. A mill was already in operation on the James River in the Virginia colony as early as 1611, constructed by technicians recruited in Hamburg.[10] (In fact, the first "patent" granted in America, by the Massachusetts colony on May 6, 1646, gave Joseph Jenks exclusive rights to operate a sawmill for fourteen years.) The sawyers were fearful that the mills would destroy their livelihood, and they often met the introduction of the water-powered sawmills with acts of violence. They tore down the first sawmill built in England, about 1633; and over one hundred years later, in 1768, five hundred London sawyers attacked and pulled down Charles Dingley's new mechanical mill in Limehouse. The same fate befell a steam-powered mill in New Orleans as late as 1803. But the threat to the sawyers increased as the mills grew steadily in number and output. The mills of the early 1700s could cut seven hundred to one thousand feet of lumber in a day operating with one up-and-down saw. Gang saws, with several saw blades in the same frame, were constructed as early as 1701. In 1769 a mill in the Mohawk Valley of New York was equipped with a gang of fourteen saws operating simultaneously. Such mills could produce well over ten thousand feet per day, and led to the eclipse of the sawyer's trade.

New methods of nail-making were also an important factor in new construction methods. Ezekiel Reed of Bridgewater, Massachusetts, was a pioneer in the development of machinery to cut and head nails, obtaining patents in 1786 and 1798. In 1795 Jacob Perkins of the same state patented his machine for cutting and heading nails. First put into operation in Amesbury, it had a capacity of two hundred thousand cut nails per day. Reed and Perkins were followed by many others who turned their ingenuity to improvements in cut-nail manufacture and perfecting machinery for even more rapid and cheaper production of the wire nail.

By 1830, the volume production of sawn lumber and the greatly reduced cost of nails made balloon-frame construction feasible. Heavy timbers with hand-shaped joints were quickly supplanted by lightweight plates, studs, and roof rafters joined with spikes and nails. The city carpenter of 1850 had no further use for the felling axe and the broad axe. Hand production of timber by sawyer and carpenter was relegated to the farm or rural areas distant from the riverside sawmills. The carpenter continued to use the hand saw, chisel, plane, drawing knife, hammer, bit brace, and other small tools throughout the age of wood, but the great axes were put aside as construction materials produced by his hand gave way more and more to materials produced by the machine.

FIGURE 92. Joinery in door construction. Model of door panel and molding of stile and rail. The small ovolo plane (*top*) shapes the edge molding of the framing members; the plow plane (*right*) forms grooves for insertion of the raised rectangular panel; the panel plane (*left*) shapes the sloping beveled edges of the panel.

V

JOINER AND CABINETMAKER

The American Mechanical Dictionary of Edward H. Knight gave the following definition of "joinery" in 1876:

As distinguished from carpentry, the art of framing the finishing work of houses; doors, windows, shutters, blinds, cupboards; hand-railing of stairs, balconies, and galleries; mantle-pieces (if of wood), the construction of permanent fittings, and the covering of all rough timber.

The carpenter is supposed to use the axe, adze, chisel, and saw, and has to make and place the wall-plates, joists, and other floor-timbers, floors, staircase, the woodwork of the roofs, wooden partitions, and those portions of the frames of windows, doors, and skylights which are built into place.

It is not surprising that the *Dictionary* does not define the craftsmen once identified as "joiner" and "cabinetmaker," for these terms had passed into general disuse by the latter part of the nineteenth century. Knight's definition of joiner describes not an artisan, but a multiple-purpose woodworking machine. Interior and exterior finishing of fine homes, which required hand craftsmanship, was performed in 1876 by the more highly skilled members of the carpenters' trade. For the ordinary finishing, any reasonably good carpenter could produce adequate results by availing himself of stock trim and moldings ready-made by machine.

In seventeenth-century England and the countries of Europe, carpentry and joinery were still relatively distinct occupations. Moxon introduced his three essays on joinery by defining it as *"an Art Manual whereby several Pieces of Wood are so fitted and joyned together* by straight lines, Squares, Miters, *or any* Bevel, *that they shall seem one intire Piece."* The French cabinetmaker Roubo writing in 1769, and English authors of works on the mechanical arts in the early 1800s, such as Thomas Martin, continued to distinguish between joinery and cabinetmaking on the one hand, and carpentry on the other. In colonial America, however, the differences between the occupations tended to be considerably less distinct. Carpenters and joiners are referred to separately in the earliest records, but there are extremely few references to cabinetmakers as such. In inventories, the individual who can be identified as having done joinery invariably possessed a larger assortment of tools and more specialized ones than the carpenter. The Wethersfield farmer Edward Veir, who died in 1645, only nine years after the Hartford plantations were settled, was likely a spare-time joiner. Among his tools was "one plough playne," the joiner's adjustable plane for grooving, which he might have used in making wainscoting. From newspaper advertisements of the following century it is apparent that a craftsman often identified

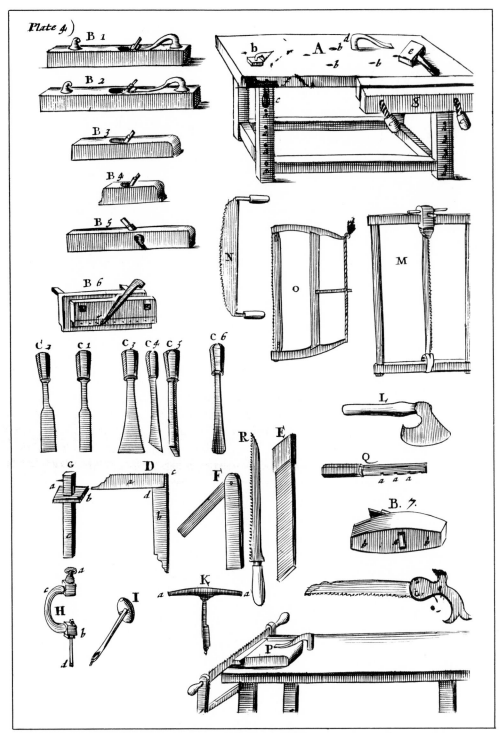

FIGURE 93. Tools of the joiner, depicted by Moxon in 1678. These complement
the tools of the carpenter (Figure 81), and include planes (B 1–7), chisels (C 1–6),
square (D), miter square (E), bevel (F), marking gauge (G), bit stock (H), and
a saw wrest for setting saw teeth (Q).

himself as "carpenter and joiner," "shop-joiner and cabinetmaker," or "cabinetmaker and chair maker." He combined closely allied crafts to which the tools and skills were basically common.

Homes built during the first years of settlement had little structural and decorative joinery. The interior walls were at first bare, with exposed posts, studs, and beams. Furniture and cabinetry were not high among the priorities; the settlers made what was necessary to meet minimum needs, and imported almost no furniture during the first century. The bedstead was the most common article of furniture. Chairs were something of a rarity, and people sat on stools or atop wooden chests. By the eighteenth century greater comfort and refinement in domestic architecture were demanded, and specialists in cabinetmaking and joinery became more numerous, particularly in the larger towns. By midcentury, when the population of the colonies stood at one million, joiners and cabinet- and furniture-makers were regularly advertising the availability of their services or articles of manufacture.

The established joiner or cabinetmaker of 1750 possessed a wide range of woodworking tools. In addition to the common bench planes, he needed a number of molding planes and special planes made for discrete applications. His tool chest contained a wooden bit stock and a set of pads with their individual bits. He used the hand saw, and the compass, tenon, and dovetail saws as well (Figure 94). He had an assortment of chisels to provide precise fitting of joints, his mortise and paring chisels small in scale in comparison to the heavy and often ponderous chisels of the carpenter and millwright. Because the cabinetmaker had to make lathe turnings for the legs, stretchers, and back posts of chairs, for table legs, and for decorative rope turnings, he would be likely to possess his own lathe chisels and gouges. He needed a calliper to measure the diameter of lathe turnings (the points were set to the measurement re-

quired and the diameter of the work was tested from time to time until it was reduced to the proper dimension). The use of shell, scroll, and other carved motifs to ornament furniture made it likely that he would also have a set of carving tools. He had squares, bevels, rules, gauges, and other measuring tools, made and finished to exacting standards, often from imported hardwoods of attractive grain and coloring. They were not infrequently produced by the artisan himself.

Rasps and files were also important tools for joiner or cabinetmaker. They used the wood rasp for preliminary rough shaping and cutting away of small surfaces, the cabinet rasp for coarse work, and the file for smoothing and fitting. Files of American manufacture, as well as sandpaper in three variations of grit, are listed in the 1838 catalog of Wm. H. Carr & Company of New York. Other than file-making by the local blacksmith and isolated small manufacturers, however, there was no quantity production of these tools in America until the middle of the nineteenth century. Sheffield, the Lancashire area of England, and Germany were the major sources of supply until that time. Before the development of practical file-cutting machinery in the 1850s all files were completely handmade; every wide tooth of the file and pointed tooth of the rasp was formed by the file-cutter by striking a small chisel with a heavy file-maker's hammer. During the Civil War, both the Philadelphia firm of Henry Disston & Sons and the Nicholson File Company of Providence began to manufacture files and rasps with file-cutting machines, and they became two of the largest producers of these tools.

Estate records list tools of two mid-eighteenth-century joiners, Jacob Leavitt of Fairfield, Connecticut, and Daniel Ballard III of Boston.[1] Leavitt owned four saws, two bit stocks, five chisels, seven turning chisels and gouges, eight files and rasps, twenty-five planes, as well as hatchets, an adz, hammer, squares, gauges, a compass, a "good bench," and other

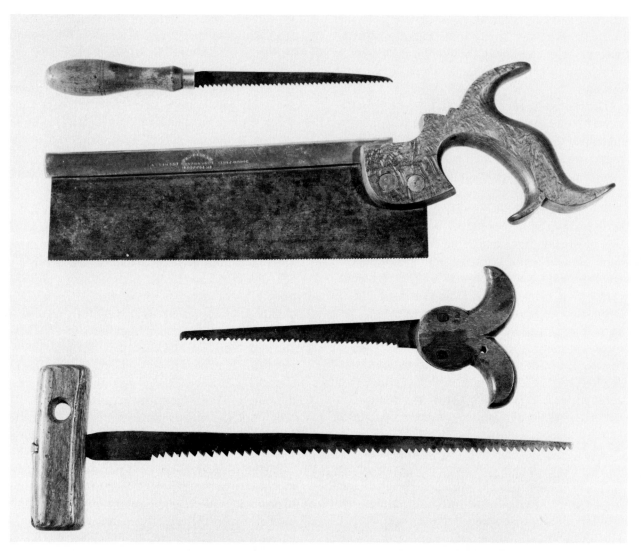

FIGURE 94. Small saws. Two narrow-blade keyhole saws and a larger compass saw for making curved saw cuts. The largest, the tenon saw, has a brass back strip to stiffen the blade. Lengths, 8⅞ to 15⅛ inches.

necessary incidentals of his trade such as a glue pot and whetstones for sharpening his tools. Some tools were probably overlooked in the inventory, for there was no mention of a mallet, drawing knife, spokeshave, or rule, all listed in the itemization of Ballard's tools, which is otherwise quite similar to Leavitt's. The twenty-five planes Leavitt owned were by no means excessive. The tool chest of the renowned New York cabinetmaker Duncan Phyfe (now dis-

played at The New-York Historical Society) contains sixty-one planes, but not the jack and jointer planes and other large bench planes, which have apparently been lost.[2]

Beginning in the late sixteenth century, and for the next two hundred years, a number of books on architectural design were published which profoundly influenced joinery and the range of woodworking tools required. Some

FIGURE 95.   Rasp and float. *Above:* This rasp came to America many years ago. Marked "*B T. BRAMALL*," it was hand cut by Thomas Brammall, filesmith of Whitehouse, Sheffield, England, who was working in 1787. Length, 14½ inches. *Below:* Float, a single-cut filing tool used by plane-makers to finish wedge slots in plane stocks. Length, 12⅝ inches.

FIGURE 96.   Cabinet rasps and files. From Nicholson File Company catalog of 1894.

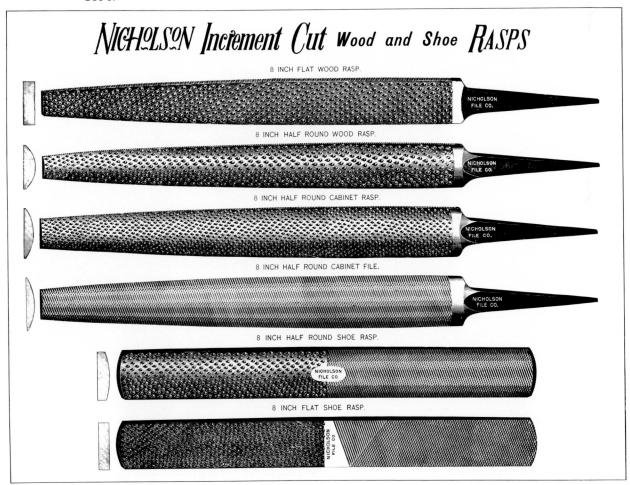

FIGURE 97.   Joiners' try squares. Rosewood handles and brass trim are characteristic of these tools for testing right angles and scribing lines perpendicular to a base. Blade of small square, 6 inches.

122

FIGURE 98. Double caliper. Cabinetmaker's tool for measuring the diameter of work as it is turned at the lathe. Length, 17¼ inches.

illustrated grandiose villas, town and country houses, and public buildings, and the common theme running through them was the "discovery" of the architectural orders of classical Greece and Rome. *I Quattro Libri dell' Architettura* by the great Italian architect Andrea Palladio, first published in 1570, was extremely influential. This work, with its plans and elevations of buildings in classical style, first appeared in an English translation in 1715. It was an obvious antecedent of *Palladio Londinensis; or, The London Art of Building*, which a Colchester carpenter, William Salmon, published in 1734. Salmon's work was designed as an architect-builder's handbook, illustrating mensuration and the classical orders. The author included one of the first price lists for work in carpentry, joinery, and other trades, gave contemporary prices of some tools and hardware, and provided a builder's dictionary. *The Builder's Director, or Bench-Mate* of the English architect Batty Langley, issued in 1751, was a similar handbook. It consisted chiefly of illustrations of Greek, Roman, and Gothic orders, with proportional scale drawings of columns, doorways, chimneypieces, and molding patterns. The architectural detail, carved in marble and stone in the days of Greece and Rome and in the villas of Palladio, was thus translated by Salmon and others into designs that would be reproduced in wood. These English pattern books were imported for sale to local builders, but an American architectural book of this type soon appeared. *The Country Builder's Assistant*, by the New England architect Asher Benjamin, published in 1797, was the first of his several practical guides for the builder which were widely studied and used. It reflected a neoclassical style inspired by the author's contact with the work of the Boston architect Charles Bulfinch. Benjamin's plan for a Federal-style meetinghouse resulted in examples that today grace many New England communities. As John Quinan has commented, "The elevation is a handsome one, calculated to

lie just within the capabilities of a country housewright. Benjamin's shrewd awareness of the limitations of his audience insured the popularity of *The Country Builder's Assistant*, and this explains, in part, the special popularity of the 'Design for a Meeting House.' "[3]

The Romans had used molding planes to copy in wood some of the decorative treatments of their stone architecture, but the tools became obsolete when the classical styles fell into disuse. However, joiners' wooden planes, each capable of shaping a different molding pattern, reappeared in the seventeenth century. They provided the means to execute in wood much of the ornamentation of the revived classical style depicted in the contemporary architectural handbooks, which replaced the more fanciful wood carving of the medieval and Renaissance periods as the principal form of interior architectural decoration.

Few American houses built in the 1600s are standing today. Of those extant, most have been modified and rebuilt so that the original construction is largely concealed. It is apparent from what does remain and from secondary evidence, such as the limitations of the tools known to have been available, that these homes had little exterior ornamentation. By the early eighteenth century, greater affluence of some individuals, the publication of architectural pattern books, and the manufacture of new kinds of tools, particularly molding planes, made it possible for the joiner to introduce greater decorative treatment of wood into construction.

A classical molding re-created with molding planes is illustrated in Figures 100 and 101. The model is scaled approximately to the proportions illustrated by Langley in his *Builder's Director*. The joiner used this type of molding in the cornice of the chimneypiece (which consisted of mantel board and its supporting moldings). He also used molding planes to form similar decorative woodwork in the architrave above both exterior and interior doorways and over windows.

MOULDINGS.

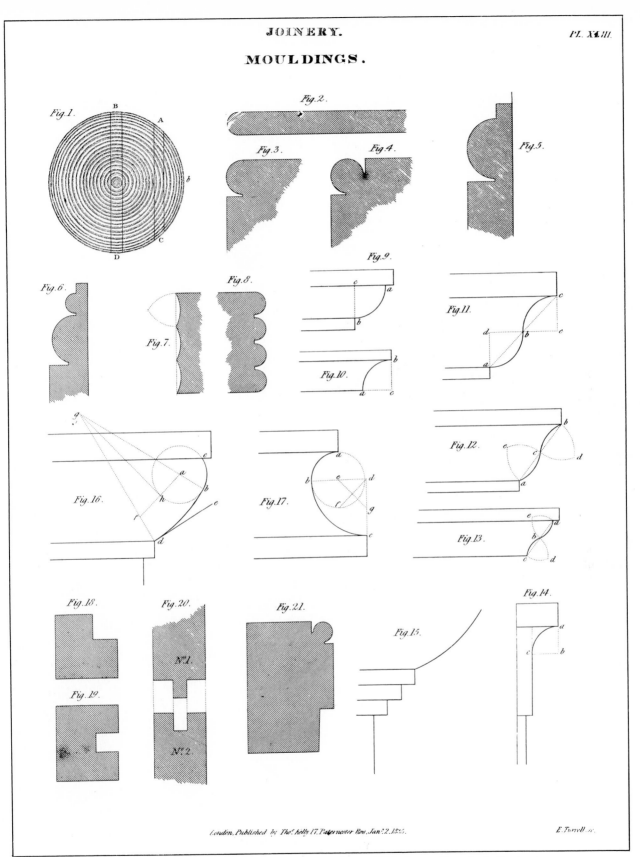

London, Published by Tho.º Kelly 17, Paternoster Row, Jan.ª 2, 1855.

E. Turrell. sc.

FIGURE 99. Moldings in joinery. From Peter Nicholson, *Practical Carpentry, Joinery, and Cabinet-Making*, London, 1846.

FIGURE 100. Joiner's moldings; a model. Reproduction of the Corinthian entablature of the temple of Antoninus and Faustina in Rome.

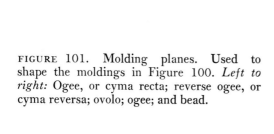

FIGURE 101. Molding planes. Used to shape the moldings in Figure 100. *Left to right:* Ogee, or cyma recta; reverse ogee, or cyma reversa; ovolo; ogee; and bead.

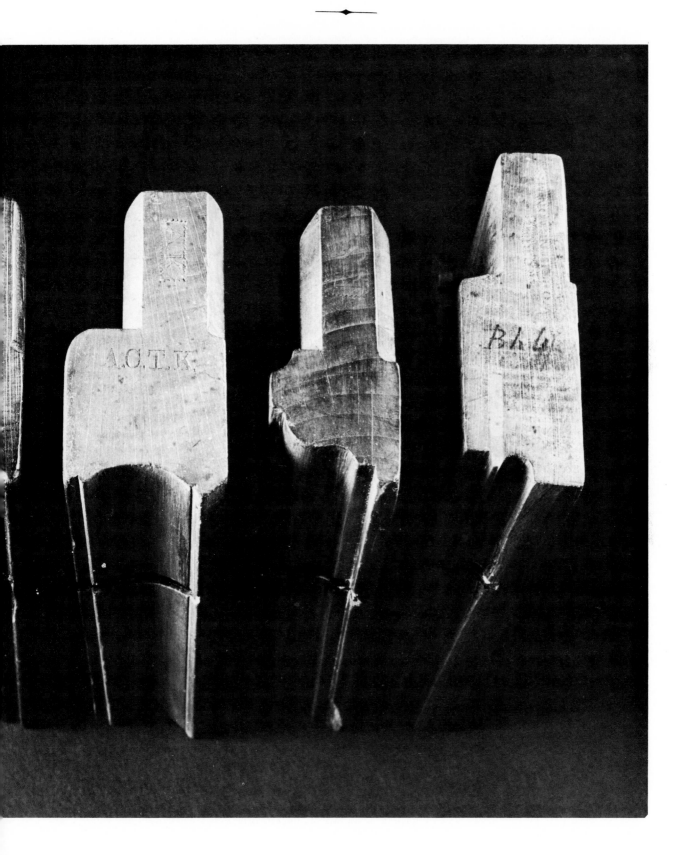

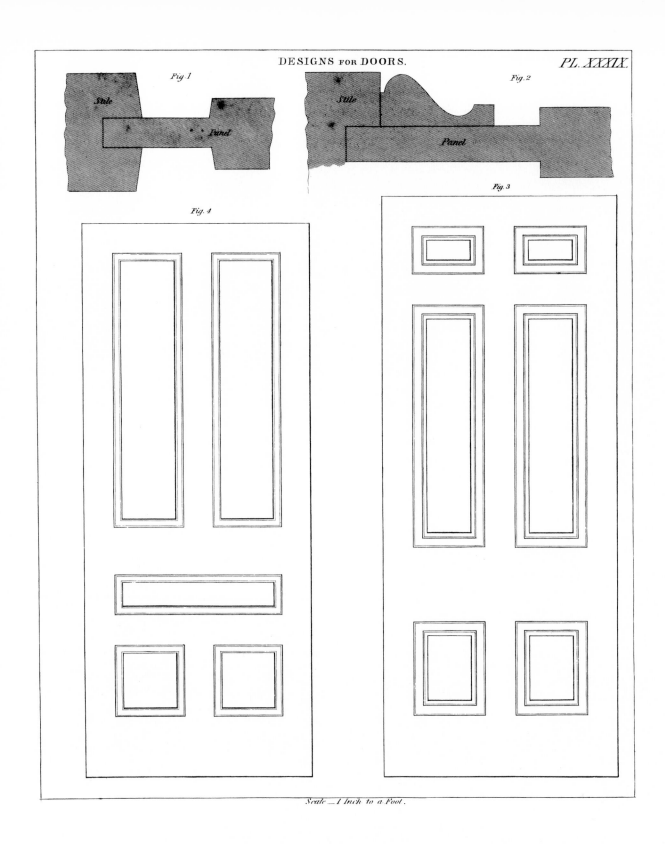

FIGURE 102. Panels and moldings for door construction. Designs from Asher Benjamin, *The Architect; or, Practical House Carpenter*, Boston, 1848.

128

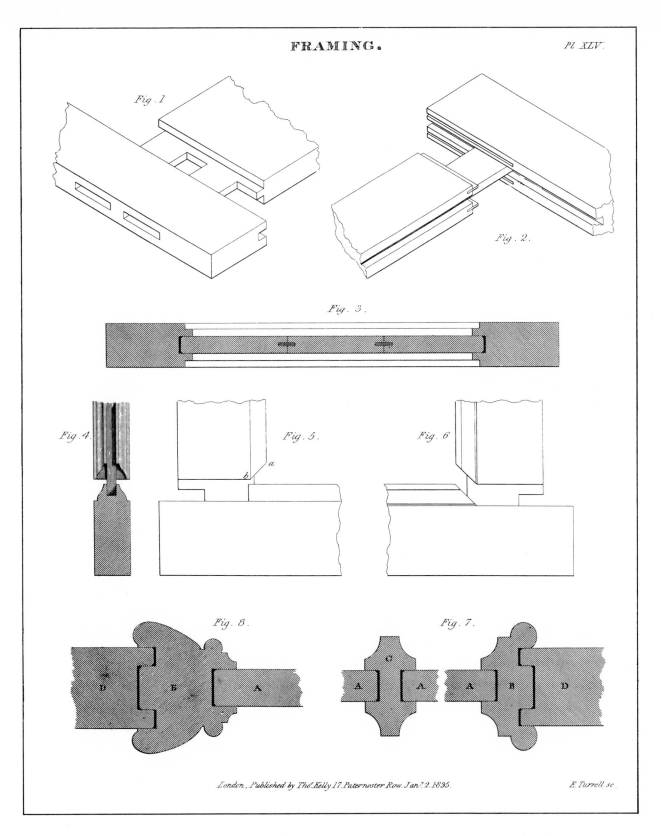

Fig. 1.

Fig. 2.

Fig. 3.

Fig. 4.

Fig. 5.

Fig. 6.

Fig. 8.

Fig. 7.

London, Published by Tho.ª Kelly 17. Paternoster Row. Jan.ʸ 2. 1835.

E. Turrell sc.

FIGURE 103. Panel framing. Mortise-and-tenon joinery for doors and solid-panel window shutters, from Nicholson's *Practical Carpentry, Joinery, and Cabinet-Making.*

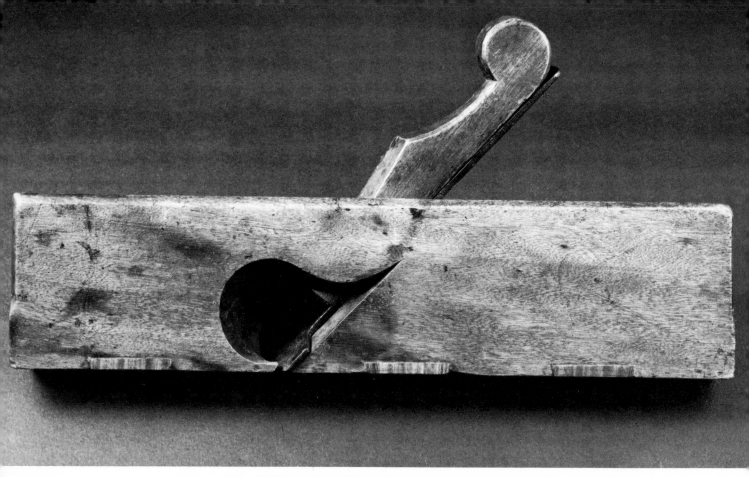

FIGURE 104. Rabbet plane. The style suggests that it was made about 1800. Owner/maker's mark, "T. HURLBIRT." Lignum vitae wear plates are dovetailed into the sole to reduce wear on the yellow birch of the stock. The gracefully curved throat of the rabbet plane provides for escapement of shavings. Length, $12^{15}\!/_{16}$ inches.

If vertical pilasters on either side of the entrance door were to be fluted, he shaped these with the round plane. To shape a cornice of composite moldings, the joiner might use several molding planes (as in Figure 101) or one large crown-molding plane (Figure 63).

In the earliest American construction either the back of the exterior clapboards or the clay filling between the studs was exposed in the interior. Plaster and wide boards, running either horizontally or vertically, soon came into use for finishing rooms. Several methods were used to treat the edges of these finishing boards. In some instances they were merely butted together with square-jointed edges. Sometimes they were planed with a halving plane to form rabbets half the thickness of the board, on alternating front and back sides, so that the edges could be overlapped. More frequently, the exposed meeting edges of the wide pine boards that extended from floor to ceiling were planed with a bead

plane, with a small ogee, or with a fancier molding plane to create a pleasing decorative effect.

As the eighteenth century progressed and the use of joinery became more extensive, the joiner revealed his skill in refinements of the wainscot. Pine or poplar wood was selected to create this framed paneling for walls. The names of the framing members, "stiles" for the vertical and "rails" for the horizontal, are the same as those which identify similar parts of doors and shutters. The joinery techniques were also similar, with mortise-and-tenon joints used at the meeting of stile and rail. Mercer noted a distinct change around 1775 in the edge moldings of rails, vertical stiles, and center upright muntins of doors.[4] Prior to that date these edges were quite consistently planed to form an unbeaded ovolo or quarter-round molding; then these simple moldings were superseded by ovolo moldings with one or two beads or ogee moldings. The joiner might insert plain flat panels

FIGURE 105. Plow plane. With sliding arms, the adjustment of the fence set by wooden thumb screws; manufactured by Hermon Chapin's Union Factory, Pine Meadow, Connecticut. Beechwood, with a boxwood wear strip inserted in the fence. Length, 9⅛ inches.

into the framing of door or wainscot, but more often the panel was "raised": a gradual, sloping bevel was formed on the four sides of the panel by a raising, or panel, plane (Figures 92, 103). The skewed cutting iron of the panel plane worked smoothly on the two edges of the panel running with the long grain of the wood. Working across the grain on the two shorter edges, however, the iron had a tendency to create rough edges on the raised rectangular field. These rough edges, typically found when an eighteenth-century panel plane was used, are evident in the pew backs in the sixteen-sided church in Richmond, Vermont.

The raised panels were fitted into grooves cut into the edges of stiles and rails with the adjustable plow plane. Moxon described this tool in 1678 and stated that the joiner should have several, so that he could shape grooves of varying widths. Around 1800 the versatility of the tool was increased by making the plane stock broader and providing a wider slot for the wedge and iron. This permitted the joiner to use cutters of varying widths. The edge-tool makers produced irons for the plow plane in widths of one-eighth to five-eighths of an inch, graduated in sixteenths.

The grooving plow was a versatile tool. Irons were available in several widths, and the depth of the groove could be regulated by means of a depth stop which was adjusted by a thumb-screw on the top or side of the tool. The plane was provided with a fence, paralleling the stock and shaped so that its depth was greater than that of the iron sole plates. The fence formed a guide which positioned the cutter iron at a constant distance from the edge of the board being planed. The distance between fence and cutter could be adjusted by means of threaded or sliding arms passing through the stock. The cabinetmaker used the plow plane or the grooving plane of a "match" pair (see page 136)

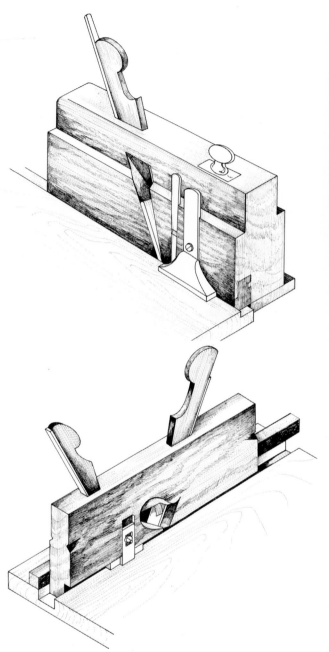

FIGURE 106. Moving fillister and dado planes. The fillister is rabbeting the edge of a board; the dado plane is forming a rectangular groove.

FIGURE 107. Moving fillister plane. Made by William Moss, who was working in Birmingham, England, in 1780. Length, 10 inches.

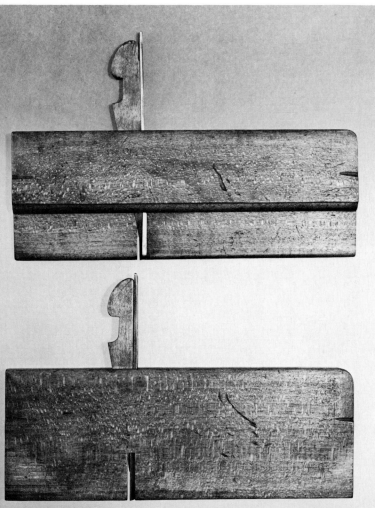

FIGURE 108. Router, or "old woman's tooth." A form of plane to remove wood in a recessed groove. A joiner or cabinetmaker made this example from beech wood. Length, 12 inches.

FIGURE 109. Side rabbet planes. The cutting edge of the iron is on the side, rather than the bottom, in this pair of planes. Length, 9½ inches.

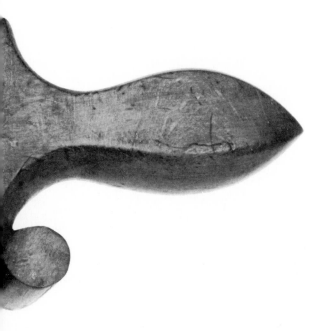

screwed to its sole. It was made with a depth stop either of wood, adjusted by a lateral screw, or with a brass fitting positioned by a thumb-screw. After a rabbet had been cut, the joiner might need to trim it to remove slight irregularities; for this purpose he used a "side" rabbet plane. A pair of these planes was needed so that the user could always work with the grain of the wood.

The dado plane combined features of both the rabbet and the fillister. It was a grooving tool, made in a series of widths from one-quarter to one inch, for planing a flat groove. The inner sides of case furniture made with fixed shelves, for example, were planed with the dado to receive the ends of the shelves. A small, ancillary plane iron, formed with two projecting knife-edged spurs, was positioned in the plane stock ahead of the plane iron. These spurs scored two parallel tracks in the wood just in advance of the cutting iron, and prevented the sides of the cutter from tearing the wood. This was essential in cross-grain work with the plane. The dado plane was made with a depth stop to keep the depth of the groove constant, but not with a fence, so it again was necessary for the cabinetmaker to affix a temporary wooden strip to his work as a guide. Timothy Tileston, a plane-maker of Boston from 1822 to 1866, produced an atypical version of the dado plane with an adjustable lateral fence similar to that of the plow plane; this obviated the need for guide strips.

when a groove was required, as on the lower inside edges of the sides and front panel of drawers of cabinet furniture; the feathered edges of the drawer bottom were fitted into these grooves. The plow plane as manufactured in the nineteenth century from wood was a handsome implement (see Plates 16, 18). The makers used brass fittings and the finest imported wood in their highest-quality models.

Rabbet, fillister, and dado planes were used by joiner and cabinetmaker. A characteristic of both the rabbet and dado plane is a graceful, sweeping curve in the throat provided for discharge of shavings. Rabbet and fillister planes were used to form a steplike rectangular recess on the edge of a board. The rabbet plane was not provided with a fence to position the tool constantly along the edge of the board being planed; the artisan had to tack a wooden guide strip to the board. The moving fillister was a type of rabbet plane with an adjustable fence

A number of additional special planes were manufactured for preparation of surfaces, fitting, and shaping in joinery and cabinetwork. The toothing plane (Figure 112) was small, no larger than the smooth plane. The upper surface of its iron was forged in a series of narrow vertical indentations. The plane stock was shaped with a vertical, or almost vertical, bed for the iron. The scraping action of the upright iron's toothed edge removed wood and lightly scored the surface. Cabinetmakers utilized the tool to prepare surfaces of core wood for the application of glue and thin veneers, and also to

plane the stubborn end grain of hardwoods. Many types of joinery and furniture-making required beveled edges. For this work, planemakers produced a chamfer plane that incorporated an adjustable block on the sole to regulate the depth of the bevel.

The surfaces of cabinetwork tended to retain slight irregularities, even when carefully leveled with the smooth plane. To eliminate such imperfections on both flat and curved surfaces, the joiner and cabinetmaker used scraping tools.

Some of these were simply small, hand-held pieces of thin steel, the thickness of a hand-saw blade, with straight or irregularly curved edges. A scraper with flat, upright blade, slightly burred along the leading edge, was used to prepare the surfaces of desks and tables. Some scrapers, such as those devised by Leonard Bailey, were manufactured with a cast-iron stock, adjustable blade, and handle characteristic of an iron plane. Another style, made of wood, resembled a spokeshave. In both types

FIGURE 110. Rogers's miter planer. The cast-iron plane with a 4-inch skewed cutter slides in an iron track for precision planing of angle cuts, as in picture framing. Patented in 1882, and made by the Langdon Mitre Box Co., Millers Falls, Massachusetts. Length of base, 29⅜ inches; plane, 23 inches.

FIGURE 111. Chamfer plane. Depth of cut, and width of the chamfer, can be regulated by an adjustable wooden fitting in the throat of the tool. Length, 7 inches.

FIGURE 112. Toothing plane. The vertical iron functions with a scraping action to surface wood with a wild grain, such as birdseye maple, and also to prepare flat surfaces for applying veneer. Length, 7½ inches.

the blade was set at or near ninety degrees to the work surface.

Cabinetmaker and joiner alike used match planes. These tools were made in pairs, one to form the tongue and the other the groove used in joining the edges of two boards. These tools were made in the same length as molding planes and in widths for use on boards three-eighths to one inch thick. For thicker boards, a pair of plank match planes were the proper tools. These were thirteen to fourteen inches long, and most were fitted with screw arms and an adjustable fence. The joiner set the fences so that the tongue and corresponding groove would be centered on the edges of the boards irrespective of their thickness.

Window sash was made by the joiner, or by a specialty joiner known as a sash-maker. There were planes to shape the stiles and rails of the sash frame, and to form the more delicate vertical mullions and horizontal transoms into which the glass panes were set. For stair building there were also special tools. Balusters, turned on the lathe, supported a handrail, or banister, whose top was molded with a banister plane. If a stairway curved, this section of the banister was shaped with a handrail shave. This tool held an iron whose cutting profile corresponded to that of the banister plane used for the straight portions.

Although the guild structure in the trades, still operative in England, was almost nonexistent in colonial America, there were exceptions, such as the Carpenters' Company of the City and County of Philadelphia, which was organized in 1724. The "Articles" for the conduct of its business, published in 1786 in its rule book, were patterned after those of a typical English joiners' company. The plates in the book illustrating examples and techniques of joinery, together with the rules for measuring and pricing all types of material and labor, are a valuable source of information on the trade in eighteenth-century America. The scope of "carpentry" in this document, it should be noted, included house joinery.

FIGURE 113. Cabinetmaker's scraper. The craftsman who made this tool for his own use was not satisfied with straight, characterless handles, and shaped these in gentle curves. A two-piece bone wear plate is inserted in the walnut stock. Length, 12 inches.

FIGURE 114. Handmade sash plane. The width of window-sash framing to be molded by the tool can be varied by adjustment of the two sections of the split stock. Length, 8¼ inches.

FIGURE 115. Stanley "Bailey" no. 12 adjustable veneer scraper. "Suitable for veneer, cabinet, and floor scraping," its mechanical design is derived from the first patent issued to Leonard Bailey, August 7, 1855, for a "plane-scraper." Length of handle, 11 inches.

FIGURE 116. Miter box. This one is of cast iron, but the wooden miter box, which serves the same function, was described by Moxon in 1678 and is still used by carpenters. The saw can be set at any angle up to 90 degrees for miter sawing of moldings. Length of box, 12 inches; saw, 17⅛ inches.

In 1835 the House Carpenters and Joiners of Cincinnati, Ohio, published their first price book. As revised in 1844, it included the usual cost per foot for timbers, wainscoting, stair railings, and other details of house construction. Its publication was very late for a book that still priced a house built by the traditional post-and-beam method and that suggested a differentiation of the trades of carpenter and joiner. (Balloon framing and platform, or "western," framing, using light two-inch lumber, had, in fact, almost completely replaced the traditional timber frame construction by this date.) The book notes that "if posts, plates, ties, sills and studs are of hard wood, add ¼."[5] Soft wood — pine and spruce — was by that time the usual wood for framing timbers in the ordinary dwelling, and oak therefore carried a premium price.

The skill of the joiner is nowhere more evident than in the hand-fashioned wainscot, doorways, staircases, chimneypieces, and the cupboards of American domestic architecture of the eighteenth and first quarter of the nineteenth centuries. A fine sense of proportion and scale was an essential attribute of the joiner, however skillful he might be in the techniques of his trade. He could and did have handbooks as guides for design and construction, and the necessary tools, but these could merely supplement the critical eye, the knowing hand, and the sum of his experience. This was especially true of the cabinetmaker in the application of the tools of his trade in furniture-making.

Cabinetmaking flourished in eighteenth-century Boston, New York, and Philadelphia. Fine furniture found a ready market among wealthy merchants in the north, and also in the larger cities of the southern colonies. Shipments to the well-to-do planters and traders of the West Indies were frequent, and ship captains occasionally bought furniture on speculation from the cabinetmakers to sell in the Indies. Newport, Rhode Island, was another major furniture-making center. There, from about 1740, the Townsend and Goddard families, Quaker craftsmen of the highest order, created some of the finest examples of colonial cabinetmaking. They were responsible for the creation of the distinctive shell-carved, block-front style (the drawer fronts of their desks, secretaries, and other case pieces have two projecting vertical panels flanking a central recessed panel). Chairs, desks, and tables of Rhode Island manufacture were regularly shipped out of the ports of Providence and Newport. In the twelve years from 1783 through 1794, over four hundred and eighty desks were shipped from Providence to southern coastal ports and almost two hundred to the West Indies; in the same period about two thousand seven hundred chairs were exported, chiefly to other colonial ports.[6]

The cabinetmaker, as well as the joiner, had sources for fashionable English designs available in books such as *The Gentleman and Cabinet-Maker's Director* of Thomas Chippendale, published in 1754, and in imported furniture that could be copied. To these he brought his own creative capacity to develop distinctive American styles. Knowledge of his materials and skill in the use of his tools were requirements of those superior cabinetmakers whose secretaries, desks, chests, and other furniture stand in our museums today attesting to their craftsmanship and art.

FIGURE 117. Caulking implements. Caulking mallet of live oak, with iron rings. Length, 15 inches. Two irons for caulking the seams of ships.

VI

SHIPWRIGHT

SHIPBUILDING IN America, as in all maritime countries, was based on the needs of trade and the fishing industry. The northeastern coast of America, for a century prior to the first attempts to establish permanent plantations, was regularly fished by the French, Dutch, English, and Portuguese. At first the fleets fished the waters off the coast of Newfoundland, but in the early years of the seventeenth century the area was expanded to include the offshore waters of Maine and Massachusetts. Cod was plentiful there as well as in the deeper waters of the Grand Banks. Fishermen brought their catch to the mainland where it was dried on racks, or salted, and packed, before they sailed back to the English and European ports. Pre-settlement exploration of the coast had also revealed its economic potential as a locale for cutting timber, trapping beaver and otter, and trading in fur with the native Indians. It was in a large measure the wish to establish permanent bases for fishing and the fur trade that led the English to send out the first contingents of settlers. Although religious separatism motivated the early immigrants, the financial backing was supplied by "adventurers" who expected to profit from an investment.

Shipbuilding in the new land preceded the settlement of New England. In 1607 fishermen of the Plymouth Company set up temporary quarters at the mouth of the Kennebec River. There, under the direction of a London shipwright, they constructed a ship of thirty tons. The vessel sailed the coast collecting furs to be carried back to England. In later years it made several ocean crossings and one voyage south to Jamestown carrying salt cod. Once the New England settlements were permanent, ship carpenters were in great demand for the construction of ships for transportation, fishing, and exploration of the surrounding coastline and waterways.

In his history of the plantation at Plymouth, Governor Bradford recorded that a ship carpenter was sent out from England in 1624. This individual, who has remained nameless, was a skilled craftsman, "thought to be the fittest . . . in the land." He exercised diligence as a master, according to Bradford, took not even an hour's time from his work, and built two good, strong shallops and a lighter. He had completed the timbers for two ketches when he "fell into a fever in the hot season of the year" and died.[1] The shallops were open boats not more than thirty feet long, with one mast, suitable for light coastal shipping and offshore fishing. The

FIGURE 118. (*Opposite.*) Lipped adz. With edges of the blade turned for dubbing the planking of ships. Made by Collins & Co. Length of head, 11 inches; width of blade, 5 inches; length of pin poll, 2¾ inches.

FIGURE 119. (*Above.*) Witchet. Rounding plane for circumference shaping, with many applications, such as shaping treenails, pegs, and handles for agricultural and woodworking tools. Length, 10 inches.

FIGURE 120. Wooden boat framing. Fishing boat under construction in the bayous of Alabama in 1977. The futtocks forming each frame section are fastened today with iron bolts, rather than treenails, but the framing retains the shipbuilding principles of the past.

lighter was a flat-bottom, open barge for harbor transportation, and the two ketches left unfinished would have been two-masted seagoing vessels with closed decks. One of the two shallops mentioned sailed late in the following year, 1625, with a load of corn to the Kennebec River and returned after making a successful trade for seven hundred pounds of beaver skins and other furs.

In spite of a high mortality rate, the population of the colonies rose rapidly within the first twenty years to over twenty-six thousand. Fishing boats for local food supply were essential, and so were freighting vessels for movement of goods, as roads were barely passable or nonexistent. Shipwrights were needed. Robert Molton was the chief of a group of six shipwrights sent to America by the Massachusetts Bay Company from London in April of 1629. Tools, pitch and tar for preserving rigging and wood, oakum for caulking, and cordage and sails were sent the following month. Molton was made responsible for storing and preparing an inventory of the provisions and tools after his arrival.[2]

The importance of shipbuilding and the fishing industry to the Colony is indicated by an order of May 23, 1639, in which the Bay Company exempted fishermen and ship carpenters from the military training required of others as preparation against Indian attacks.

The General Court of the Company in October 1641 took note that "the country is nowe in hand with the building of ships, which is a business of great importance for the common good, & therefore sutable care is to bee taken that it bee well performed, according to the commendable course of England, & other places." The owners of every ship under construction were required to appoint an overseer to be responsible for bringing any unsatisfactory work to the attention of the master ship carpenter. Failure to comply with his recommendations was a serious matter. He was authorized in such instances to request the governor, his deputy, or two magistrates to appoint two expert ship carpenters to serve as "viewers" and make further inspections to see that sound timbers were being used and that the construction met accepted standards.[3] The Connecticut Colony in May of 1673 adopted a law for the maintenance of standards similar to the earlier Massachusetts statute. It included an immediate penalty of ten pounds for failure of the owner to seek the appointment, before construction was started, of an able person to survey the carpenters and their work.

Only sketchy records exist of the numbers and types of vessels built during the first decades and through the seventeenth century, but they attest to the importance of shipbuilding. Registry of ships did not start in the Massachusetts Bay Colony until 1689. The registry was primarily for merchant vessels of over thirty tons; the smaller fishing boats, ferries, and coastal freighters were excluded. The extensive trade with the West Indies and among

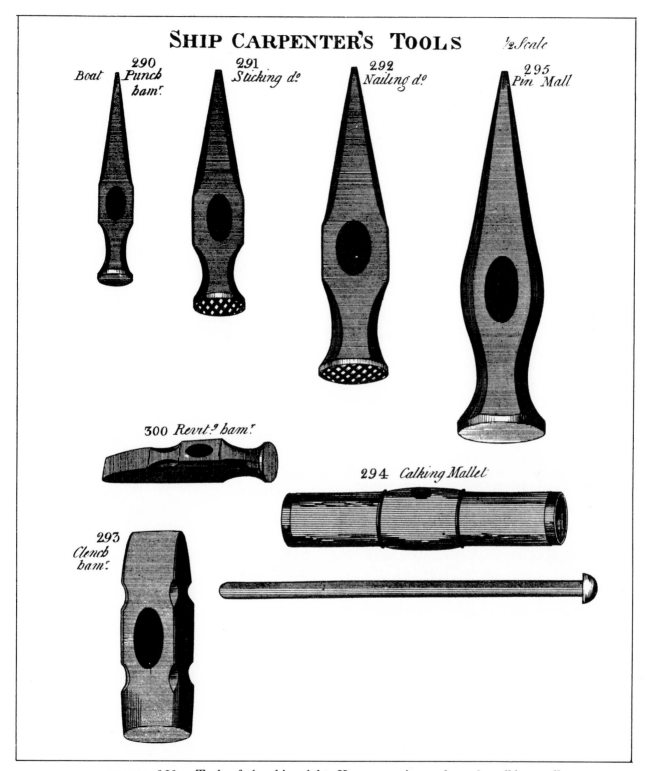

FIGURE 121. Tools of the shipwright. Hammers, pin mauls, and caulking mallet are among the tools of this illustration from Smith's *Key*.

147

the colonies was often carried on with surprisingly small ships of even less than thirty-ton capacity. The Boston Port Register of 1698 to 1714 shows over one thousand three hundred sloops, pinks, ketches, brigantines, barks, and ships built in America; of these, the great majority were owned in Boston. One hundred sixty-nine were constructed for owners in the British Isles. The record is unquestionably incomplete, but it does afford a good indication of the progress of colonial shipbuilding. Merchant-vessel construction was centered at this time in the Bay Colony, with Boston, Scituate, Cambridge, Charlestown, and Salem the towns with the most active yards. The Boston port book also records the presence of ships constructed in Kittery, Maine, and in all the colonies of the eastern seaboard as far south as Virginia.[4] Governor Leete of Connecticut made an inventory of vessels owned in that colony in 1680: New London was the home port of seven of the total of twenty-seven, which did not include small fishing and oyster boats; five were owned in New Haven.[5]

Shipbuilders in New England had access to

plentiful supplies of native timber throughout the eighteenth century. White oak was selected for keels and major framing members, for planking, rails, and decks. Locust was used for treenails. The masts and spars were made from white pine. Pine ultimately took the place of oak for deck construction. William Douglass, in his mid-eighteenth-century history of the British settlements in North America, called attention to early use of mahogany in shipbuilding, as well as its use in cabinetmaking and joinery. The West Indies trade with New England provided the source for this quality wood used in finishing the joinery of ship cabins. In the later era of the finely built schooner, clipper ship, and packet boat, the interior joinery was of the highest standard of craftsmanship and elegance, with carving and paneling of mahogany and other choice woods.

Native timber for foreign shipbuilding was a substantial item of export in the seventeenth and eighteenth centuries, as was the construction and sale of completed ships to England, France, and other European countries. Timber was very scarce in those countries, depleted by use of wood over the centuries for fuel, shipbuilding, general construction, and the manufacture of wood products. Colonel David Dunbar, Surveyor General of His Majesty's Woods, in a report to the Duke of Newcastle in 1729, told of a trip along the Maine coast and of boarding at Casco Bay a mast ship from New Hampshire then making ready to sail to England with a load of masts for the Royal Navy. Agents of the British Crown regularly searched the forests of Maine and New Hampshire, selecting choice pines to be felled for masts. These trees were identified by an arrow blazed on the bark, and were later exported for use in the Royal Navy shipyards. Portsmouth, New Hampshire, and

FIGURE 122. Smooth plane. Made by a ship joiner for his own use from lignum vitae. Length, 8¼ inches.

FIGURE 123. Spar planes. The radius of the soles varies in these tools for finish planing of ship spars. Lengths, 6⅝ inches to 7 inches.

Falmouth, Maine, were major ports for the export of mast timbers to England.

Even in America, however, the supply of timber for shipbuilding was not limitless. In 1791 Alexander Hamilton, with commendable foresight, called the attention of Congress to the export of wood, and warned of the necessity "to commence, and systematically to pursue, measures for the preservation of their stock."[6] Hamilton's warning was prophetic, for within a few short years the supply of New England white oak suitable for shipbuilding was completely exhausted. The Delaware, Maryland, and Virginia seacoast peninsulas then provided the northern shipyards with timber. The builders sent lightweight wooden framing patterns for ships' hulls, known as molds, south to these states. Lumbermen felled the great white oaks, and carpenters working in the forests sawed, shaped, and pegged together the sections to make each frame. The parts of the frames were marked, disassembled, and shipped to the northern yards. In these years, pitch pine was supplanting oak not only for decks, but for some of the framing of the ships and for the outer planking as well.

149

The hand tools of seventeenth-century American shipwrights were much the same as those used by English shipwrights of the previous century, and, moreover, they changed little throughout the entire period of American shipbuilding in wood. The published records of the Connecticut Colony include two inventories of the shipwright William Lotham, one made in March 1645 as a part of his will, and the second on September 27 of that year, following his death.[7] Lotham's goods included three thousand five hundred planks and six thousand treenails, "a barrell and three quarters of tarre and pitch, lying att the waterside . . . a boate of tenn tun . . . an anker, a grapnell, maine-saile and foresaile . . . one hundred nynty three pickes" (spikes?), an auger and a drawing knife. Lotham had an assortment of other tools and nautical gear: caulking irons to drive oakum into the seams of the planking to make the ship watertight; "some heads for clinke work" (small sledgehammers for clinching, or turning back, the points of spikes); a scraper; a "brest wimble," or bit stock; an axe and iron wedge; a pair of pincers; two hammers; a gimlet, a file, a mallet, and a gouge; a narrow chisel; two oars; two "setting poles" to pole his skiff in shallow waters; an eel spear; a pocket compass; and a skiff with two oars. Curiously absent from the lists are the shipwright's adz and saws of any kind. Perhaps hand saws and the adz were in a chest which the appraisers mentioned without noting its contents.

There was so little significant functional alteration during the period of shipbuilding in wood that the artisan of 1870 would have been at home using Lotham's tools. The radical changes came only with new methods of construction, the introduction of power machinery that made much hand work obsolete, and the development of the iron ship.

In framing the ship, the shipwright used the adz constantly. The futtocks (timbers that were pieced together and fastened with treenails to form each skeletal rib frame), the keel, the stem and stern posts, the planking, the rudder, and the masts were all trimmed with the ship carpenter's adz. The compound lateral and vertical curvature of the hull made fitting the planking an exacting task. Particular skill with the adz was required in shaping the outer and inner surfaces of the frames so that each plank, made more supple by first steaming it in a long wooden box, would fit flush against the curved frame. The adz was again used in "dubbing" the planking, smoothing the irregularities in the surface where one plank butted against another.

As for augers, the common, T-handle variety was used in shipbuilding, but also a specially adapted kind. An iron shank, up to six feet or more in length, was welded to the short shank of the auger received from the manufacturer; the elongated shank allowed the worker to bore holes in places hard to reach. An important use of the auger was to bore holes in planking the sides of a vessel: in each ship, each of the hundreds or even thousands of these holes was fitted with a treenail to hold the planking in place.

Caulking was a critical process to assure that both hull and deck would be watertight. Striking a caulking iron with an iron-bound caulking mallet, the caulker forced strands of cotton into the seams between planks (Figure 117). Over the course of cotton, he drove in "threads" of oakum — untwisted strands of hemp often from old cordage. The blades of caulking irons were made in a variety of thicknesses, with the edge either straight or shaped with a slight arc to fit the configuration and thickness of the seams.

The first step of the shipwright in mast-making was the selection of a straight tree trunk, usually pine, of proper length. From a center line, the taper of the mast was laid out and the timber squared with an axe. The axe was used again further to reduce the bulk of the mast, bringing it to an octagonal shape. The shipwright then used his mast shave, a large drawing knife, to bring the timber to its final round shape and taper. Spars for large ships were made in the same way, and finished with spar planes.

Finishing above and below decks was the

PLATES 21, 22. Four stages in the raising of a barrel. Miniature casks made by a cooper then working at Whitbread & Co., Ltd., brewers, of London.

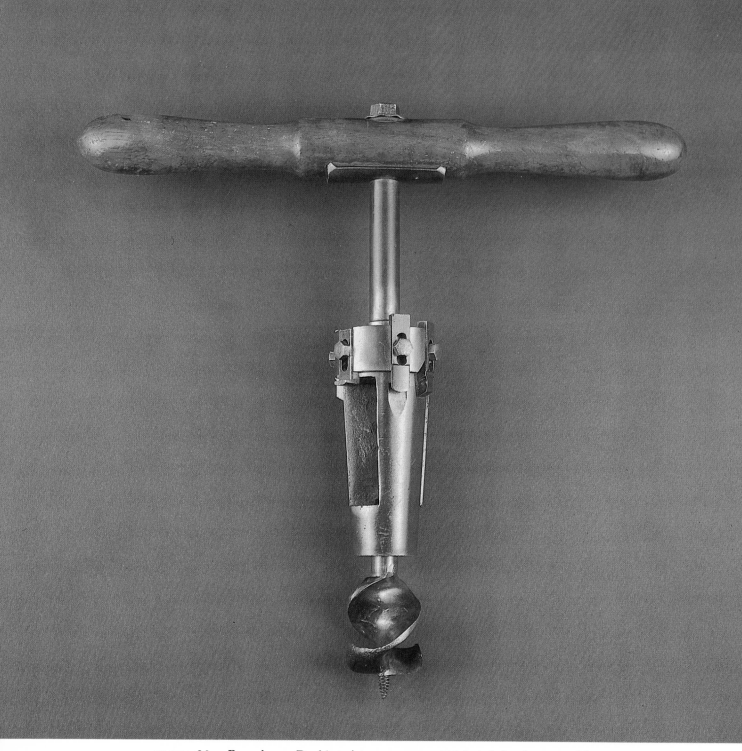

PLATE 23.   Bung borer. Double-twist auger cuts a 2⅛ inch hole; the tapered borer forms a bunghole up to 2¾ inches in diameter for fitting the bung. Length, 17⅞ inches.

PLATE 24.   Cooper's croze and howel. Cast-brass fittings complement the bird's-eye maple stocks of these tools. *Above:* Croze, for cutting notch in the chime into which the barrel head is pressed. Length, 14½ inches. *Below:* Howel, for smoothing the chime of a barrel. Length, 16¼ inches.

PLATE 25.   .Apple wood sun plane, flagging and raising iron, and chamfer knife. Tools of the tight barrel cooper. Lengths: plane, 13¾ inches; flagging iron, 18¼ inches; chamfer knife, 14 inches.

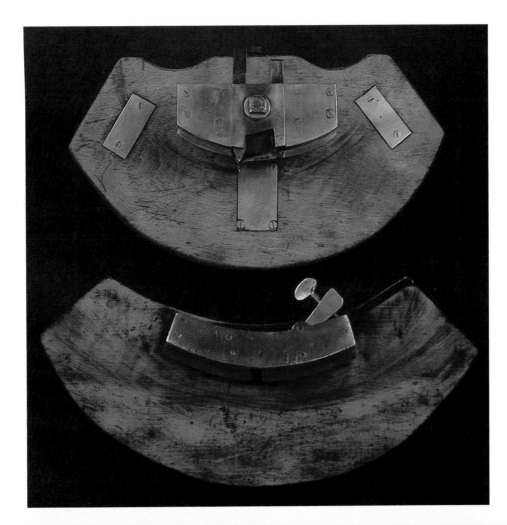

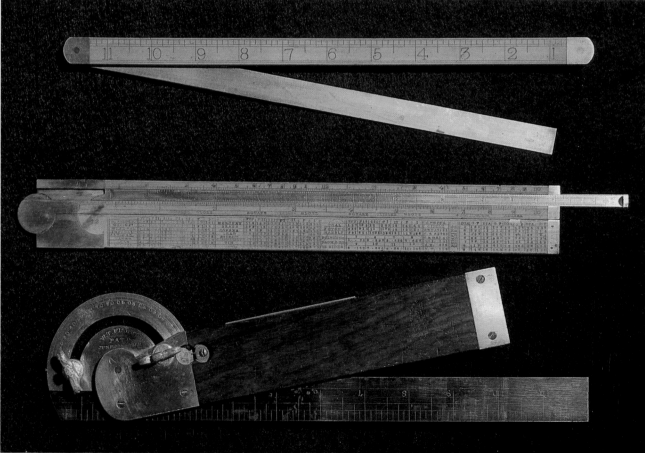

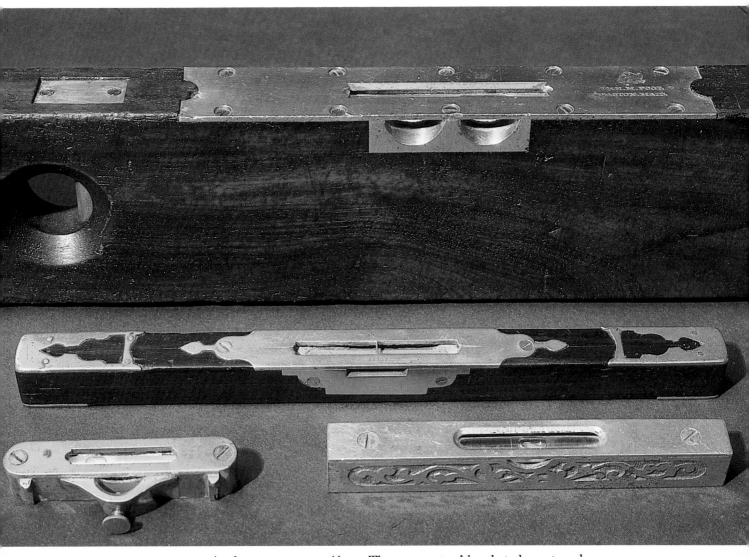

PLATE 26.   Angle measurement. *Above:* Three carpenters' bevels to lay out angles. *Below:* Mahogany joiner's square with 16¼ inch blade.

PLATE 27.   Instruments of measurement. *Top:* Shipwright's one-foot bevel in box-wood, made by E. A. Stearns of Brattleboro, Vermont, about 1850. *Center:* Two-foot, two-fold, arch-joint, boxwood rule. Stearns no. 2, with Gunter's line and calculating data. *Bottom:* Protractor, inclinometer with level, square, and bevel in one combination tool, made by Disston & Morss, Philadelphia. Length, 12⅜ inches.

PLATE 28.   Leveling tools. A spirit level and plumb (*top*) by J. & H. M. Pool of Easton, Massachusetts. *Below:* Ebony- and brass-trimmed 11-inch level made in Scotland; a small pocket level, designed to fit on the blade of a carpenter's square; and a 6-inch Stanley level in cast iron with floral decoration.

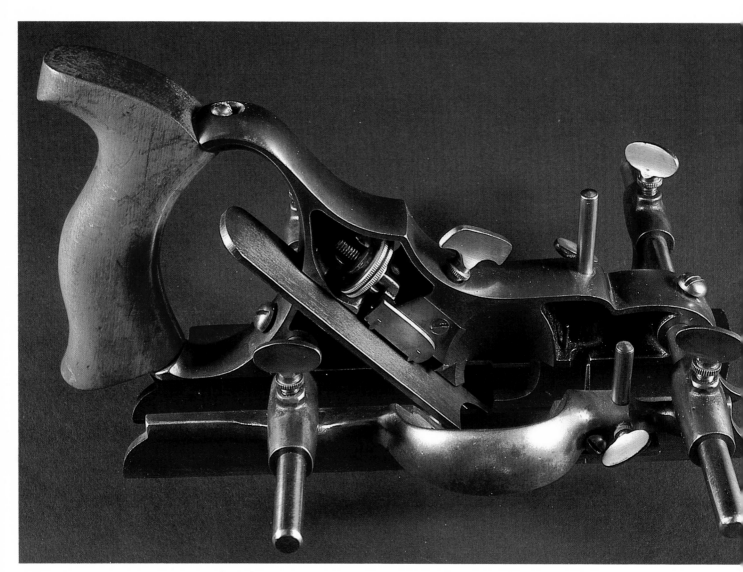

PLATE 29. Combination plane. Manufacture of Jacob Siegley of New York City on 1881 patent, in which the machinist's economy of design is apparent. But for the handle and a small fence, wood has given way to new materials and manufacturing technology. Length, 9¾ inches.

PLATE 30. Charles G. Miller's combination plow and fillister plane, made by the Stanley Rule and Level Company. The casting bears a floral design not inconsistent with tastes in decoration during the Victorian era in which it was produced. Length, 10⁵⁄₁₆ inches.

PLATE 31. Scribing the line. Mortise and marking gauges to mark construction lines in carpentry and joinery. Lengths, 7¼ to 10 inches.

province of the ship joiner, a specialist crafts-man who made the detail woodwork of com-panionways, fittings, deck houses, and cabins. He used an assortment of tools comparable to those of his counterpart on land, and often found employment ashore in joinery of houses when shipbuilding was slack.

The employment of shipwrights was subject to the vicissitudes of naval warfare, embargoes, and the general economic conditions both in America and in the countries she traded with. In the period before the American Revolution the demand for shipwrights was high. Many came from England to contribute their skills to the growing industry in the colonies. So many left England in the early 1700s that the English shipbuilders tried unsuccessfully to influence the Board of Trade to restrict emigration. Dur-ing the Revolution, shipbuilding rapidly de-clined, and in the postwar era the British Navigation Act prevented the sale of American-built ships in England. (The restrictions of the same Navigation Act also forced merchants to seek new markets. A Baltimore-built ship sailed into the harbor at Canton, China, in 1785 to open American trade in tea, chinaware, silk, and other products of the Orient.) The shipwrights of the port of Philadelphia in particular com-plained about the effects of the Act, petitioning Congress in 1789 to take action. The war be-tween the English and French also brought hardship to the American shipping industry. Nearly eight hundred American vessels were seized at sea and in English and French ports. The United States was forced in December of 1807 to place an embargo prohibiting her ships from engaging in foreign trade. It was a period of unemployment in the shipyards, and thirty thousand seamen were idle. The contraction of shipbuilding and foreign commerce had a favorable side, nevertheless. It provided a stimu-lus to industrial expansion.

Following the War of 1812 there was a revival in shipbuilding: ships were needed to replace those lost or captured during the pre-vious decade, to accommodate passengers in a swelling tide of immigration, and to meet re-quirements of increased foreign trade. A new class of ship, the packet boat, was put into ser-vice. Most of these were built in New York shipyards for fast transatlantic passenger, mail, and light freight transportation. The industry prospered, and Maine took the lead, building almost one-third of all shipping produced be-tween 1830 and 1860. The Houghton and Sewall yards in Bath completed ninety-five ships in the years 1840 to 1882. In Massachu-setts, Donald McKay, at East Boston, con-structed fifty-nine ships between 1845 and 1869. The clipper ships of the twenty years from 1840 to 1860 secured for America the major share of American carrying trade. They could regularly cover three hundred nautical miles in a day, and the *Flying Cloud*, a McKay-built ship, once sailed four hundred and twenty-seven miles in one day.

The final impetus for the construction of wooden sailing vessels came with the discovery of gold in California in 1848; freight and pas-senger traffic increased sharply for a time, but by 1857 the unusual demand had been fully met. The wooden ship, built chiefly by the hand labor of the shipwright, gave way in the follow-ing decades to the ship of iron.

FIGURE 124.   Tools of the wheelwright. *Clockwise from top:* Tapered auger for hub boring; spoke dog, for fitting spokes to felloes; traveler, to measure a length of iron for the tire; wheeler's mortise chisel, to mortise the hub for setting spokes; felloe pattern, from which outlines of the felloes were traced for sawing to shape; and (*center*) felloe saw.

VII
WHEELWRIGHT AND CARRIAGE-MAKER

THE FARM wagons, carriages, and coaches built in the two centuries prior to the American Revolution were relatively sophisticated forms of the animal-drawn wheeled vehicle. Little significant change took place in their functional design or in the methods of construction, other than those associated with woodworking and metal fabricating processes, up to the time of their virtual disappearance in the twentieth century.

The basic components of the wheeled vehicle are a platform or body and attached axle, wheels, and double shafts, or a single pole, for harnessing one or more draft animals. In prehistoric vehicles, wheels did not rotate freely on the axle; rather they were pinned to the axle, and wheels and axle rotated together between vertical slots or pins on the underside of the platform. The transition to wheels rotating freely on axle spindles took place more than three thousand years ago.

There is some likelihood that the earliest wheels were round cross sections cut from tree trunks. A more satisfactory form of wheel, used by the Sumerians in the third millennium B.C., was constructed from three, or less frequently two, parallel sections of plank pegged together and sawn to form the circular shape. In eastern Anatolia, primitive cart wheels of this construction are still in use today. A subsequent major

development, about 2000 B.C., was the lighter, spoked wheel with wooden rim, replacing the solid wooden wheel. A magnificent ninth century B.C. alabaster relief of King Assurnasirpal II depicts the Assyrian ruler on a lion hunt, riding in a chariot with spoked wheels. Actual examples of even earlier wheels, of the fourteenth century B.C., came to light with the discovery in 1922 of the tomb of Tutankhamen; several wheels and the parts of four chariots were preserved in the tomb. The wheels were made with wooden hubs from which projected six spokes, attached to the rim by metal fittings. The rims were made of several pieces of wood, like the felloes of the nineteenth-century wagon- or carriage-wheel.

Although there is ample archaeological evidence of the antiquity of the crafts of chariot-builder and wheelwright, early writings on the subject are rare. An important record of the Chinese Chou Dynasty (ca. 1122–256 B.C.) is the *Khao Kung Chi*, or "Artificers' Record," which reveals the state of technological development two to three thousand years ago. It states: "Woodwork includes the making of wheels, chariot-bodies, bows, pikestaffs, house-building, cart-making and cabinet-making with valuable woods."[1] Hubs with holes to accommodate tapered axle spindles, spokes, jointed felloes to form the circumference of the wheel,

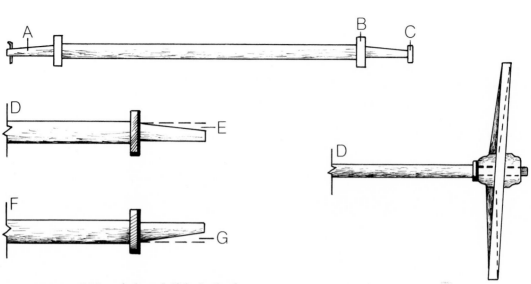

FIGURE 125.   Axle and dished wheel.
A. Axle spindle   E. "Swing," downward bending of spindle
B. Butting ring   F. Axle viewed from above
C. Linch pin, or nut   G. "Gather," forward inclination of spindle
D. Axle viewed from front
Bending the axle spindle downward and forward allows each spoke of a dished wheel
to be perpendicular at the moment it is directly below the axle.

metal tires, bronze hub bearings which were greased, and the "dished" wheel were all in use in the Orient over two thousand years ago.

The dished wheel was probably the last fundamental development in wheel construction. Its design provided increased strength to withstand sideways pressure when a vehicle encountered rutted roads or was driven on a sidehill. In Europe, this type of wheel was used from the late Renaissance period, and in America it was routinely used for wagons and carriages during the eighteenth and nineteenth centuries. Structurally, the dished wheel is an almost flat cone; its spokes radiate from hub to rim at a slight angle. This wheel was normally used on a tapered spindle (the conical hub bearing at each end of the axle). When used on a horizontal tapered spindle, a wheel has a tendency to work against and wear out the retaining linchpin or nút holding the hub on the axle; the tapered spindle is, in effect, a continuous inclined plane, and the wheel rotation creates a lateral force toward the smaller end of the taper.

To overcome this problem, the blacksmith bent the iron axle spindles off center in two directions: downward, so that the lower edge of the spindle was parallel to the ground; and forward, so that the leading edge was at approximately a right angle to the direction of the vehicle's travel. When the spindle was at the proper angle (measured with an axle gauge), the spoke directly beneath the axle would be perpendicular to the ground as the dished wheel revolved, offering the maximum support for the weight of the vehicle and its load. This same principle of adjustment of the attitude of the spindles and wheels — called swing and gather by the wheelwright — are applied today to set the camber and toe-in of automobile wheels.

The manufacture of and trade in wagons, coaches, and carriages developed slowly in the early years of America. Land vehicles were not a significant factor, either for communication or transportation, in the period prior to the American Revolution. Colonial settlements were

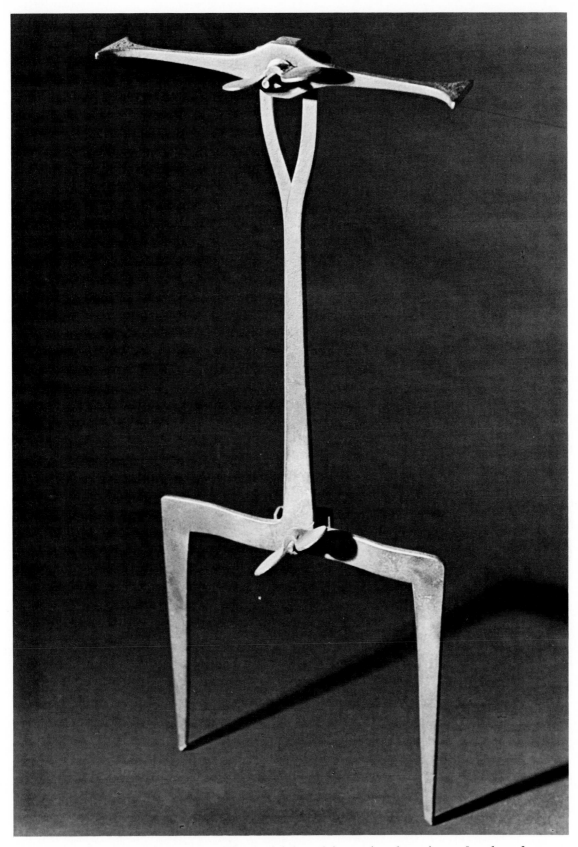

FIGURE 126.   Axle gauge. Wheelwright's tool for setting the swing and gather of
axle spindles. Length, 26½ inches.

155

chiefly confined to the Atlantic seaboard, to towns with good port access; and there were few inland towns that could not be reached by river and lake. Thus most travel and freighting was by ship or raft, on coastal routes, lakes, or navigable rivers, so the early development of a system of roads was unnecessary.

Those roads that did exist before 1780 in many instances followed centuries-old Indian trails. They were neither crowned nor ditched, and could accommodate little but slow, hazardous travel on horseback. As late as 1789, a Connecticut Congregational missionary reported with some bitterness on his travels in upper New England: "From Burlington Bay, I set out alone and unaccompanied to Shelburn through the wilderness on the Lake Champlain — next to no rode — mud up to my horse's belly — roots thick as they could be. . . . My horse nearly gave out, excessively worried by the bad travelling."[2]

Coach- and carriage-building was confined to Boston, New York, Philadelphia, and other cities, chiefly for sale and use within town. Eighteenth-century newspapers printed few advertisements of coach-builders and wheelwrights. William Cooper announced in the *New-York Gazette or the Weekly Post-Boy* of April 21, 1763, that he had set up a coach- and harness-making shop in Elizabeth Town, New Jersey. He had recently arrived from Long Acre, the center of the coach-making trade in London. In the *New-York Mercury* of September 28, 1767, Elkanah and William Deane informed the public of their ability to perform coach-building work "in the best Manner." Their plain coach and harness for two horses was priced at £165, and with "Livery Lace, and fringed Seat, Cloth and richly painted and finished," the price went up an additional thirty-five pounds.[3] It is apparent that the automotive "optional accessory group" of today, including wheel covers, "cordovan leather" upholstery, and vinyl roof, was carried over directly from the selling practices of the coach-maker.

In 1756 the highway trip from New York to Philadelphia took three days; it is understandable that freight shipments between these towns were made by the considerably cheaper and faster water passage. The first regular stagecoach route between Boston and Providence, major towns only forty miles apart, was not established until 1767. The records of tonnage capacities and destinations of ships in colonial ports in 1770 reveal the significant amount of waterborne trade that took place among the colonies. Of the goods and produce outward bound from the colonies in that year, ninety-nine thousand tons were exported to Great Britain and Ireland, thirty-seven thousand tons to southern Europe and Africa, and one hundred and eight thousand tons to the West Indies. But over thirty percent of the total tonnage was shipped to ports within the colonies, and to Bermuda and the Bahamas.[4]

The American Revolution created a need for new and better roads to move troops, equipment, and provisions. Several were constructed during the war along strategic military routes by felling trees and laying the trunks across the roadbed; these were so-called corduroy roads. After the war, with independence secured and increased mobility of the growing population possible, the first major movement away from the eastern seacoast took place. Contemporary census data reveal internal relocation within the new nation. The population of Vermont, a mere five thousand in 1771, grew to eighty-five thousand in 1790, and to one hundred and fifty-four thousand by 1800. The population of Kentucky, the only other inland state at the time of the 1790 census, was seventy-three thousand in that year, and it increased to two hundred and twenty thousand by the close of the century. Connecticut and Rhode Island, however, together showed only the very modest population gain of eleven thousand between the 1790 and 1800 censuses.[5]

Highway construction and maintenance was an obligation of the towns during the pre-

Revolutionary period. Legislation often required every able-bodied male to provide his labor for a period of a few days each year to work on the town roads. The system did not meet with great success, however, for the service period often turned into a social occasion, with fun, games, and drink, rather than the hard work needed for road repair. So the farmers of southern New Hampshire and Vermont, driving their homemade wagons to the Boston market, had to follow roads that were little better than rutted pathways, and the Pennsylvania Germans, freighting produce from Lancaster County in trains of hundreds of huge Conestoga wagons, fared no better. Well before the turn of the century the expanding needs for moving farm products, and also manufactured and imported merchandise, created pressures for building more satisfactory roads and for new methods of financing their construction.

Intermittent attempts to raise money for road work by public lotteries proved unsatisfactory. The toll road system, borrowed from England, was somewhat more successful. It was introduced in Virginia in 1785, in Maryland two years later, and in Connecticut in 1792. Turnpike corporations chartered by the state legislatures built the majority of toll roads during the last decade of the eighteenth century and the first ten years of the 1800s. The corporation charters granted the financial backers the exclusive right to collect tolls over prescribed routes. The income was to create funding for maintenance of the highways and occasionally, with luck, to return a profit to the investors. For the most part, the turnpike corporations were not successful. Maintenance was relatively expensive, and shunpiking a common practice. Petitions of the citizenry to legislative bodies maintained that main roads from the agricultural interior to the seacoast market cities should be free public roads; by 1850 the majority of the toll roads had been replaced by publicly supported highways.

Burgeoning production of farm, factory, mill,

and mine spurred dramatic improvements in highways in the first forty years of the nineteenth century. By 1835 the flat-bottom and the keel boat carried freight on the Erie Canal and on a growing number of inland canals, the steamboat plied the rivers and Great Lakes, and the railroad had arrived. New inland routes with improved gravel surfaces, new ferries, and bridges opened the way for faster, more dependable, and cheaper road transportation, and for the development of commercial teamsters. The market for wagons, coaches, and other road vehicles expanded.

The construction of wheeled vehicles required the services of a number of craftsmen. The wheelwright, primarily a worker in wood, fashioned the wheels. The wagon-builder, or wainwright, constructed the undercarriage, body, and shafts for farm and freight wagons. Of course, many blacksmiths were also wheelwrights, and farmers built wagons in the earlier days and in country locations. The term "carriage" was frequently used to identify any vehicle for carrying people, including the coach, buggy, gig, surrey, phaeton, and other styles. The trades, however, distinguished between the carriage-maker, who constructed the wheeled undercarriage, and the coach-maker, who built the coach body attached to, or suspended above, the undercarriage. A parallel may be drawn between the relationship of the house carpenter and joiner and that of the carriage-builder and coach-maker: in each case the latter performed more exacting work than the former and used a considerably greater variety of hand tools. In his work on the manufacturing establishments of Great Britain, published in 1843, George Dodd described a much more refined division of labor: "The construction of a coach," he stated, "requires the aid of coach-body makers, carriage-makers, coach-smiths, coach-platers, coach-beaders, coach-carvers, coach-trimmers, coach-lace makers, coach-lamp makers, harness-makers, coach-wheelwrights, coach-painters, herald-painters, and various others whose occu-

FIGURE 127. Wheelwright's shop at Colonial Williamsburg. *Background:* Turning a hub on the great wheel lathe. *Foreground:* Wheel on the block, preparatory to boring the hub.

pations form more or less distinct branches of trade."[6] Carriage and coach work had certainly become highly specialized.

Some specialization in these crafts already existed in colonial America. An early Connecticut record tells of the request of Thomas Munson on February 9, 1651, for a portion of an untenanted lot so that he might build a home and work in New Haven.[7] Munson stated his purpose as "setting aboute makeing of wheeles." At a meeting of the townsmen, Munson's request for the site was granted on the condition that he "build a suitable house upon it, and follow the trade of makeing wheeles, for the good of the Towne, and plowes and other things for the furtherance of husbandry as he can." The grant illustrates that the wheel-

wright's work existed as a specialized trade at that early date. Undoubtedly the farmers of the town were perfectly capable of constructing their own wagon bodies, but the services of an experienced wheel-maker could well be of use. The record also indicates that Munson was experienced as a blacksmith: he could make iron plates for the mold-boards of plows. He must have made the iron bands with which the wheel hubs were bound, the curved iron strakes (sections that formed the iron tire before tires forged in one piece were used), and rivets for fastening the strakes.

Wagons for use on the farm and for bringing produce to town, and two-wheeled carts for transporting goods within town were the principal vehicles used in the first century of settle-

FIGURE 128. Wheelwright of Colonial Willamsburg. Shaping spokes with the drawing knife.

FIGURE 129. Reaming the hub. The wheelwright uses a tapered auger to shape the center hole for the axle spindle. In wheelwright's shop at Colonial Williamsburg.

FIGURE 130.   Hollow auger. Fitted to the chuck of a bit brace, the tool shapes the round tenon of a spoke. This adjustable model cuts tenons from ¼ inch to 1¼ inches in diameter. Length, 6⅞ inches.

ment. Those who made these vehicles used many tools common to the woodworking trades, and a number associated chiefly with the wheel-wright's trade. The wheelwright used the felloe saw, a typical sturdy frame saw with a narrow blade, to cut out the curved segments of the wheel rim. Each felloe, and the curvature of its arc, was made to a pattern so that the total would form a wheel of the diameter wanted. The number of felloes depended on the number of spokes, because each felloe was bored to receive two spokes radiating from the wheel hub. The wheelwright used an adz to shape the inner curvature of the felloes. Oak, ash, or hickory felloes were joined by drilling holes in the ends of each section and inserting wooden dowels. The brace used by both wheelwright and car-riage-builder often was constructed entirely of iron to withstand the rough treatment to which it was subjected. In making spokes, the wood was riven with a froe, rather than sawn, to maintain a straight grain, and thus the maxi-mum inherent strength of the oak or hickory used. The spokes were then trimmed and

shaped with drawing knife and spokeshave, or with a special shave with a concave cutting iron and sole. The rectangular spoke tenons driven into the mortises of the hub were shaped and trimmed with chisels. The outer ends of the spokes, to which the felloes were attached, were first pointed with a conical spoke pointer and then cut to the required diameter and depth with a hollow auger. The adjustable version of this versatile tool, which came into use in the nine-teenth century, could be set to shape a round spoke tenon of whatever diameter was needed, whether for the heavy wheels of a wagon or stagecoach, or for the lighter wheels of a buggy.

The wheel hub, made from elm, oak, or sometimes beech, was first turned to shape and size on a lathe. It was made from unseasoned wood so that as it dried it would tighten its grip on the tenons of the spokes. The mortises for the spokes, always even in number, were marked out on the circumference of the hub and first bored with an auger to remove the waste wood. They were then carefully shaped with chisels — flat chisels to pare the sides and a V-shaped

160

wheeler's chisel, or corner chisel, to square the corners. "Boxing" the hub was one of the final steps in its construction. A hole was bored through the center of the hub and then enlarged to a conical form with a tapered reamer. The iron box, a cone-shaped bearing, was inserted and made tight by driving small wooden wedges into the end grain of the hub. To prevent the hub from splitting, a round iron band was driven onto each end.

The wheelwright or blacksmith made and fitted an iron tire to the wheel, usually before boxing the hub. He measured the circumference of the wheel with a traveler (an iron, brass, or wooden wheel, with a circumference of up to twenty-four inches, attached to a short handle), by counting its revolutions, and transferred this measurement to a length of iron tire-stock. He hammered the flat bar into a circle on the anvil or formed the circle with a tire-bending device, welded the ends together, and heated it in a fire. The red-hot tire was then lifted from the fire, quickly driven onto the wooden rim, and doused with water to shrink it to a tight fit. Tiring was completed by fastening the tire to the rim with rivets or tire bolts.

The wagon-maker used a splitting froe to rive the strong oak members of the body. His augers and bits for boring were made from heavy stock because he worked in elm, oak, and other hard woods. Drawing knives were in constant use to shape the spindles, rails, and other parts of the frame and undercarriage. The skilled builder carefully chamfered these parts to eliminate excess weight and to lighten the appearance of the vehicle.

An excellent description of the exacting craft of coach-building is contained in the third part of Roubo's work on joinery (published in Paris, 1769–1775). The coach body required a frame, consisting of horizontal and vertical structural parts, which was enclosed with wooden panels on each side, each with a door, with panels on the front and back, and a wooden floor and roof. The roof top, and often a portion of the back

and upper sides of the body — the "upper quarters" — were covered with leather. The coach-maker's procedure was in many essentials similar to that of the shipwright. With chalk he transferred a scale drawing of the coach body to full size on a large blackboard, and from this made a series of thin wooden patterns, or molds. Using the patterns, he outlined the pillars, bottom boards, rails, roof rails, and other framing members on wood. The patterns took into account the variations in curvature and dimension at different points of cross section, just as the mold-boards of the shipwright provided the shape for each frame of a ship. Coach construction was typical joinery work, incorporating mortise-and-tenon joints and structural framing grooved for the insertion of panels. Dry ash was used in England and America for frames, although Roubo stated that elm or walnut was used for this purpose. The panels enclosing the body were of relatively green wood; poplar or linden or, if left uncovered, mahogany was used. The panels had to be sawn and planed sufficiently thin so they could be bent when subjected to heat to conform to the curvature of the body.

The chief difference between the tools used in coach making and in joinery is in the design of planes. The structural parts of the coach body were all shaped in curves of varying degree; these curvatures were either in plan or in elevation, or both. The planes therefore could not have the long, flat soles of standard joiners' tools, but rather had to be no more than a few inches long. There were three general types of coach-makers' planes: rabbeting, grooving, and molding planes. Rabbet planes were used on the front and center vertical pillars of the coach frame to form a recess so that the hinged doors of the coach would fit flush with the body when closed. The rabbet plane was also used on other parts so that the edges of flat panels, some of which were nailed onto the frame, would not protrude. Grooving plow planes were used primarily for shaping the recesses in pillars, in

FIGURE 131. Travelers. Cast-brass, forged-iron, and sheet-iron travelers, which measure wheel circumference by a count of rotations. The measurement is then transferred to flat iron tire-stock, for measuring the length required for the tire. Diameters, approximately 7½, 7½, and 5¾ inches.

the horizontal middle rails, and in the bottom side boards of the frame. The thin panels that enclosed the lower quarters of the body were fitted into the grooves in the pillars, rails, and bottom boards. The coach-maker's plow plane was essentially similar to the joiner's, but was of smaller scale and its iron sole plates did not extend the entire length of the sole.

The coach-maker's molding planes were also shorter than the joiner's; they were four to six inches long, while the joiner's were nine and one-

half inches. Bead, cove, ogee, and other moldings were used to decorate the pillars and roof edges of the coach, and for interior trim. The curved surfaces made it impossible for the coach-maker to use molding planes, even his foreshortened types, for smoothing, cutting, or grooving in some places. Where the radius of a curved surface was too small to accommodate his planes, he worked with routers and tools known as "carriage maker's molding tools." The panel router was a substitute for the groov-

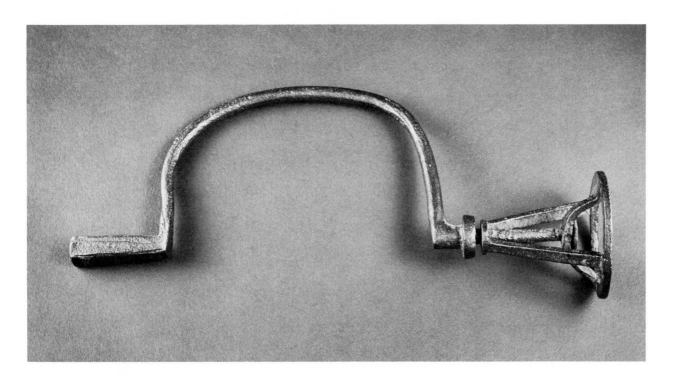

FIGURE 132. Cage-head brace. Carriage-builder's boring tool, blacksmith-made in iron. Length, 15⅝ inches.

FIGURE 133. Wheelwright's clamp. Presumably hooks clamped under the felloe; the screw held a rivet in place through tire and felloe while the opposite rivet end was flattened. Length, fully opened, 15 inches.

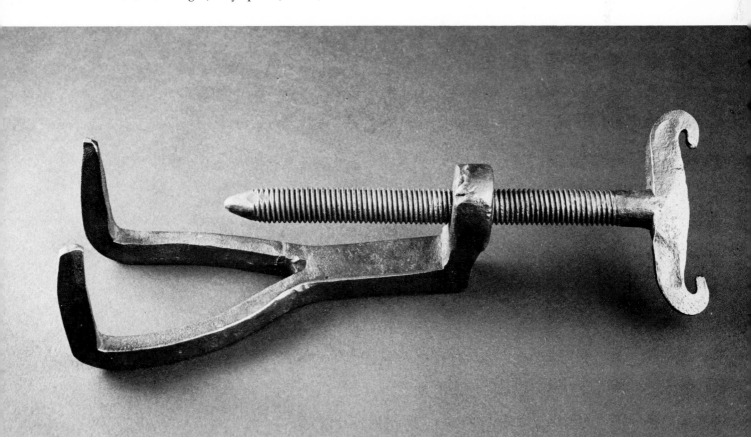

FIGURE 134. Grooving tools of the carriage-maker. *Above:* Carriage-body drawing knife. Length, 19½ inches. *Below:* Grooving router, with metal fence, to groove curved pillars and rails of the coach frame for inserting body panels. Length, 13⅛ inches.

FIGURE 135. Coach-maker's planes. Four rabbet planes, those at top and right with curved soles. The three lower tools, with soles projecting laterally, are known as "T rabbet planes." Lengths, 4½ to 6½ inches.

ing plow plane. It was similar to a spokeshave and had only a small working surface around the cutting iron; it could therefore make curved grooves. The molding tools, such as the hand beader, were also superficially similar to spokeshaves. They were often made of iron or brass in the nineteenth century.

Another tool used exclusively by the coach-maker was the grooving saw. It had a short blade, about six or eight inches long and one-eighth inch thick, which was set into a wooden stock. It was used to form grooves or fillets on the middle rails of the inside of the coach. The edges of the interior leather or cloth upholstery were fitted into the grooves, and then covered with a decorative molding.

Carriage-building for domestic and foreign trade became a significant industry in the United States during the nineteenth century. The twenty percent *ad valorem* duty that was placed in 1800 on imported coaches, chariots, phaetons, or other carriages and their parts certainly contributed to the encouragement of production. In a report on the "Arts and Manufactures" of the United States transmitted to the Senate by President Madison in 1814, Tench Coxe took note of the export of carriages to South American countries.[8] Three years earlier, Treasury Secretary Albert Gallatin had commented on the "high degree of perfection" of American manufactures in wood, including "coaches and carriages, either for pleasure or transportation."[9] At midcentury the *Census of Manufactures*, always an incomplete statement at best, showed almost four thousand carriage-building establishments, employing more than twenty-seven thousand workers. The value of their products was stated as above twenty-six million dollars.

New Haven became a center for carriage-building with the establishment of James Brewster's factory in 1810. The city had twelve carriage factories by 1831, manufacturing vehicles valued at a half-million dollars. The use of machinery came early in these New Haven factories; in 1838 one plant was operating saws and lathes with an eight-horsepower steam engine and was producing the wooden parts for building carriages on a volume basis. The city also became a center for independent manufacturers of carriage fittings: two-thirds of the forged iron axles, over one-third of the carriage springs, and a goodly share of the carriage bolts made in the United States in 1860 came from the factories of New Haven and its surrounding towns. The New Haven Wheel Company was one of the largest makers of carriage wheels, and supplied local and a nation-wide trade.

But like the other woodworking trades, carriage- and wagon-making followed the migration of the population away from the east coast. Factories appeared at Elkhart, Indiana, Chicago, Saint Louis, and points west and south, and New Haven's carriage industry languished. Well before the end of the century the industry itself had greatly changed. The wheelwright became a lathe hand and the coach-maker an assembler of machine-made parts. The use of hand tools for carriage-making was relegated to small shops in country towns.

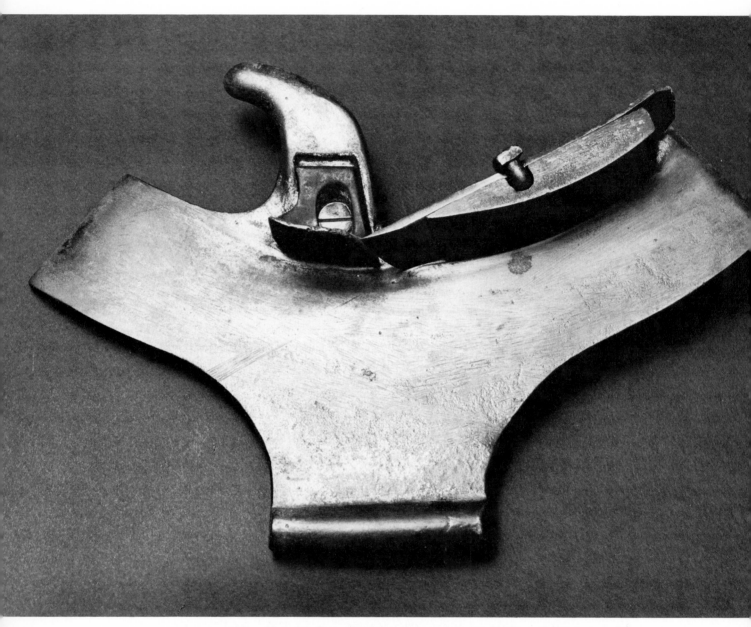

FIGURE 136. Howel in iron. Cast-iron tool made by C. F. Close in Rochester, New York, a major center of coopers' tool manufacturing. Length, 13¼ inches.

VIII
COOPER

THE EARLIEST vessels to hold liquids and dry produce were probably made by potters, at first from sun-dried clay and later from kiln-fired pottery. It is likely that the first wooden vessels were sections of tree trunks, hollowed out with stone implements or by burning and removing the charred wood. Much more sophisticated wooden containers, with separate staves and hoops, are depicted as early as the third millennium B.C. in Egyptian tomb carvings and paintings. Staved barrels and other forms of cooperage for storage and transport were in common use in Europe and England from Roman to modern times.

The cooper's trade assumed an importance in early America which is difficult to appreciate in the present era of metal barrels, cardboard containers, specialized transportation for frozen food, and frost-free home refrigerators. Most containers — from small water and liquor kegs to fifty-four-gallon hogsheads — were produced by the cooper. Cooperage was needed to pack and store fish taken on the six hundred miles of the Grand Banks in the 1600s; fish was a critical source of food for the colonists. Coopers also provided the barrels to pack the dried fish for export; it was a major product in the early colonial trade with England, and in later seventeenth-century trade with the West Indies and Spain. In addition to barrels, forms of cooper-

age required in the early years of settlement included pails for carrying water, well buckets, and tubs for household washing and for the blacksmith to quench his iron.

Governor Bradford of Plymouth has given a unique account of the work of one group of colonial coopers. Less than four years after the settlement of the plantation, coopers located on the Maine coast to make casks for shipping salt cod were called upon to perform an unusual salvage operation. In April 1624, the pinnace *Little James*, sailing to Maine on a fishing voyage, was caught in a violent storm off Damariscove Island. The ship's side was stove in with a hole "as a horse and cart might have gone in," the seas carried her to deeper water, and she sank. The masters of several English ships fishing in the area offered to supervise salvage and provide workmen. Governor Bradford agreed, and sent additional men along with a quantity of beaver skins for payment. In Bradford's words, "they got coopers to trim I know not how many tun of cask, and being made tight and fastened to her at low water, they buoyed her up; and then with many hands hauled her on shore." Ship carpenters then completed the repairs. The *Little James* was saved — but came to a hapless end in the following year. Loaded with cod and furs from the plantation, she was almost within sight of port at Plym-

167

outh, England, when she was captured by a "Turks' man of war" and taken to the north African coast, where her master and crew were sold as slaves.[1]

Almost all goods shipped from England to the colonies were packed in casks of various sizes, each size denoted by a name. The bill of lading of the ship *Lion*, which sailed to Charlestown in 1631 carrying casks made by the English cooper Edward Clarke, listed hogsheads, half hogsheads, barrels, half barrels, firkins, pipes, and runlets, all casks of differing capacities. Cheese, butter, oil, vinegar, and other food and equipment was packed in these containers. John Winthrop, Jr., then in England, paid Clarke's bill in the amount of eight pounds, eighteen shillings, for the cooperage. It covered almost fifty casks, and included separate charges for putting in the heads, setting hoops, and hooping and nailing.[2]

Coopers, as well as other artisans, were still sorely needed in the colonies. Governor Winthrop wrote to his son three years later, in 1634, asking that three artisans, including a cooper, be sent. And when the younger Winthrop later came to America and moved to a new settlement at Pequot (New London), Connecticut, Governor Winthrop wrote to him in 1647 saying that he would try to locate a cooper for that new settlement.

In 1648 the coopers of Boston and neighboring Charlestown petitioned the General Court of the Bay Company for permission to form a Company of Coopers. The court took note of numerous complaints of defective cooperage and on October 18 authorized the coopers to meet, select a master and two wardens, four to six assistants, a clerk, a gauger, a sealer, a packer, and other officers to be elected annually. The company was granted authority to control the trade by calling to the attention of the court any person "who shall use the arte or trade of a cowper, or any part thereof, not being approved by the officers of said cowpers to be a sufficient workman." The court would then have power to

prohibit such a person from working. Not unmindful of the potential for monopoly and price-fixing they were granting, the court stipulated that "no unlawful combination be made . . . for inhancing the prices of caske or wages, whereby either our owne people or strangers may suffer."[3] In December 1648, in what was one of the relatively few instances of the establishment of a guild or company on the English pattern in America, the coopers drew up their Orders of the Company.

In the light of the critical dependence of the colonies on sound cooperage, it is understandable that standards for coopers' work were frequently a matter of legislative action. "Casck and Cooper" warranted a separate section in the first code adopted by the General Court of Connecticut, in 1650. "Every Casck commonly called Barrills or halfe hogsheads shall contain twenty eight gallons wine measure, and other vessells proportionable," read the law. Gaugers were to verify the proper size of each cask by measurement and affix their marks on the casks. Each cooper was required to burn a "distinct Brandmarke on his owne Casck."[4] Nine years earlier, this same court had established rules for the dimensions of pipe staves and for their price: they were to be four feet, four inches in length, at least half an inch thick (not counting any outer sapwood), no narrower than four inches, and be sold at five pounds per thousand.[5]

In addition to establishing standards for cooperage, the courts also regulated the season and location for cutting timber, whether "for pipestaves or some other merchantable commodity." Land within the boundaries of a royal grant, outside of town lots and acreage specifically allocated to settlers, was colony property. In 1649, the Connecticut Court authorized the owners of a ship docked at Wethersfield on the Connecticut River to cut a shipload of pipe staves as freight for a trading voyage, provided they felled the timber outside the limits of any town.[6]

The cooper's trade expanded through the

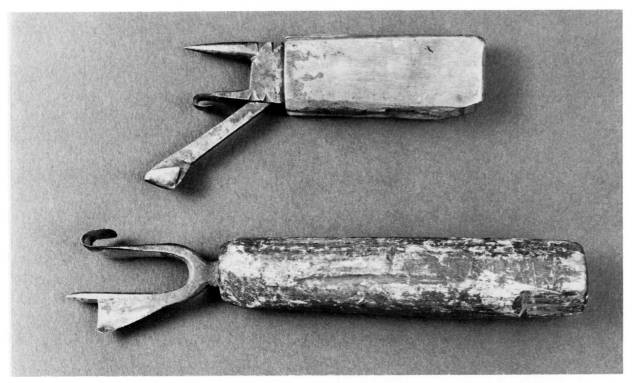

FIGURE 137. Race knives. Tool of the carpenter, cooper, lumberman, and ship-wright. The hooked blade scored timbers, staves and heading, or logs with identification marks.

eighteenth century to fill the needs for transporting goods in a growing foreign market and a home market that was moving south and west. To maintain honest measure and standards of workmanship in a trade that was producing a major commodity in domestic and foreign commerce, the colonies continued to legislate proper assize of casks and the cooper's identification of his product. In addition to completed casks, the production of staves and heading for the use of coopers in the West Indies, England, and Ireland continued to be a growing business. In 1770 more than twenty million staves and heading were exported, and in the two years from October 1791 to September 1793 fifty-nine million were shipped.[7] The exports from Virginia exceeded those of both Massachusetts and New York: white oak, the wood primarily used for wine, whiskey, and molasses barrels, was fast disappearing in the north.

Apart from export cooperage, hundreds of thousands of barrels were required each year for storing and transporting within the colonies tobacco, sperm oil, meat, fish, and countless other products. The cooper was part of the crew of every whaling ship. He made his barrels chiefly ashore; he set up and numbered the staves and heads consecutively with a race knife. He then disassembled each cask, which greatly decreased the storage space required on shipboard. The broken-down casks, known as "shooks," were reassembled at sea and filled with oil as the whales were caught and the blubber rendered.

Three distinct categories of coopers were engaged in the trade: white coopers, slack-barrel coopers, and tight-barrel coopers. The white cooper, occasionally referred to as a "cedar cooper," made round containers with straight sides that usually tapered to a smaller diameter

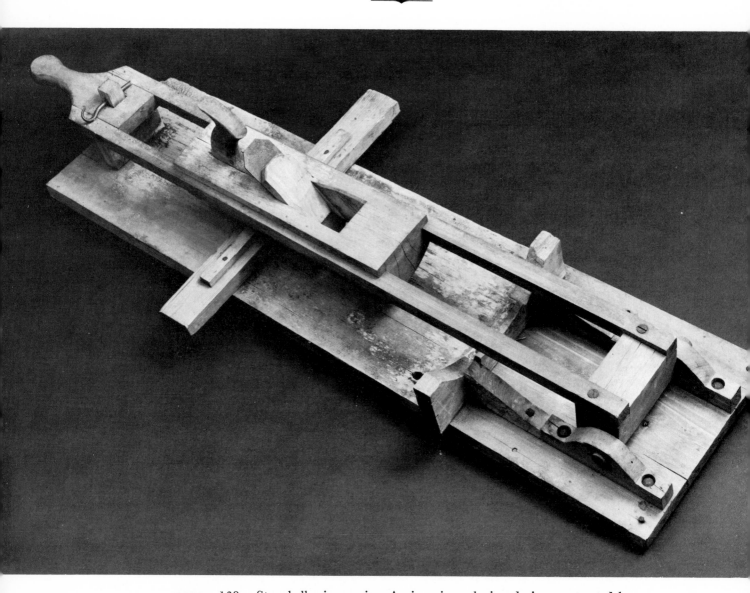

FIGURE 138. Stave-hollowing engine. An ingenious planing device constructed by a white cooper of New Hampshire. The plane slides on two tracks to hollow bucket staves placed below the plane. Length overall, 44¼ inches; length of plane, 14 inches.

at the bottom. They ranged in size from small piggins — which had one stave extended at the top to serve as a handle, for drinking or ladling — to huge gathering tubs made for sugaring, which held hundreds of gallons of sap from maple trees. Large round tubs to water cattle in the barn; maple sap buckets; cheese, butter, and lard tubs; churns; milk pails; well buckets — these were all the work of the white cooper. His material was normally the white pine, cedar, or other soft wood, and either wooden or metal hoops were used to band the staves together.

The slack-barrel, or dry, cooper made containers in the typical barrel shape, smaller in diameter at the heads and wider at the bilge, or middle, section. His manufacturing techniques were similar to those of the tight-barrel cooper, but his workmanship could be less exacting and setting up of his type of cask was less difficult. Ordinarily, a slack barrel was not made to contain liquid. It was therefore fashioned from cheaper grades of pine, beech, or poplar, and of lighter stock than the oak used for the tight barrel. The staves had to be bent to form the barrel shape, but because they were thin they could be bent without wetting and heating. Flour, meat, apples, potatoes, sugar, nails and other hardware items, and other manufactured and agricultural products were shipped in slack barrels.

The tight-barrel, or wet, cooper, had to have the greatest skill. This cooper constructed his casks from white or red oak, with stout staves and heads, and made them watertight. Shaping and butting the staves and forming and fitting the heads of tight barrels, made to hold wine, molasses, cider, oil, beer, vinegar, or whiskey, called for considerable expertise. In addition to making his casks from wood that was stronger than that used by other coopers, the tight-barrel cooper used considerably thicker staves and heading. His barrels not only had to withstand rough handling as they were rolled about in shipment, but had to remain tight under pressure from expansion of the contents because of heat or fermentation.

"Barrel" has become a general term applicable to any container made with staves and a center bulge. It was originally a specific term for a variable size of cask made to contain from twenty-eight to as many as fifty gallons. The tun was the largest standard cask, with a capacity of two hundred and fifty-two gallons. A pipe was a half tun, or one hundred and twenty-six gallons. Capacity of the hogshead varied from sixty-three to two hundred and six gallons, and of the tierce from sixty-four to eighty-four gallons. Firkins and runlets, small containers, held about eleven and eighteen gallons, respectively. The capacities of these containers varied widely, however, by local custom and by the type of liquid or dry produce with which they were filled. These variations, as well as deviations from standard in any handmade product, made it necessary to measure capacity and content by gauging rods and mathematical computation when casks were filled and shipped.

Among the cooper's tools the drawing knife appears in a variety of forms: a type with a straight blade and handles in the same plane as the blade was used by the cooper in "backing" to shape the outer, convex surfaces of barrel staves and to trim material for hoops. A drawing knife in the author's collection, with handles slightly angled down from the plane of the blade, is marked as a "heading" knife, to chamfer the circumference of barrel heads. The cooper used still another form of the tool for shaping staves, one with a concave blade to shave the inside surfaces. These hollowing knives were manufactured in three degrees of curvature: regular, medium, and full curve, the first for the largest casks, and the full curve for the staves of small containers.

The cooper prepared the wood for his staves and heading well before he intended to use it. The green wood selected for staves was either split with froe and froe club or sawn to rough dimension, and then stacked to cure. The cooper first "listed" a stave — formed a taper at each end — by chopping away a portion of each side with the cooper's axe. He next shaped the staves

FIGURE 139. Straight froe. Knife-edge split wood for barrel staves. This froe with small 9-inch blade was probably a white cooper's tool.

with drawing knives while seated at a shaving horse, a low workbench designed so he could grip the stave by foot pressure and hold it in position for shaving. The worker then planed the edges of the staves on a long plane known as a cooper's jointer. This tool was set up in an inverted position so that the cutting iron was on the upper surface. The cooper pushed the stave over the stationary plane, rather than moving the plane over the wood. For wet cooperage it was essential that he obtain the proper bevel for jointing the edges of staves. Jointing to the correct angle was a matter of "eye" and experience for the accomplished cooper, although stave gauges were available to test both the curvature and bevel. To raise a barrel (illustrated in Plates 21 and 22), the staves were set upright in a temporary raising hoop set on the ground; the work was then inverted and additional temporary hoops were driven on to hold the staves in place. The cooper then wet the inner sides of the staves and "fired" the cask, by setting it over a small fire kindled on the ground in a round open framework of iron hoops known as a "cresset." Moisture and heat made the staves more pliable so that they could be bent inward at the head ends. With the hammer end of his adz and a wooden hoop driver, the cooper hammered temporary truss hoops made from tough hickory down and around the staves to force them into final position. Once the stave ends were brought tightly together and as the permanent hoops of wood or iron were fitted in position, the truss hoops were removed.

The barrel head was made in sections, bored on the edges and doweled together. To obtain the correct size of head for his cask, the cooper used a compass or dividers. He adjusted the legs of the dividers, by trial and error, until the

FIGURE 140.   Mallet and curved froe. White coopers used a curved froe for bucket staves. Froe clubs were rough tools, quickly fashioned and soon discarded.

FIGURE 141. Shaving horse at Colonial Williamsburg cooper's shop, for holding staves in rounding and backing.

FIGURE 142. Cooper's jointer. Plane is set with sole and cutting iron facing up. The cooper edge-planes staves by passing the wood over the iron. Manufactured jointers were made in lengths from three to six feet. In the shop at Colonial Williamsburg.

FIGURE 143. Cooper's shop. Cooperage in progress at Old Sturbridge Village, Sturbridge, Massachusetts.

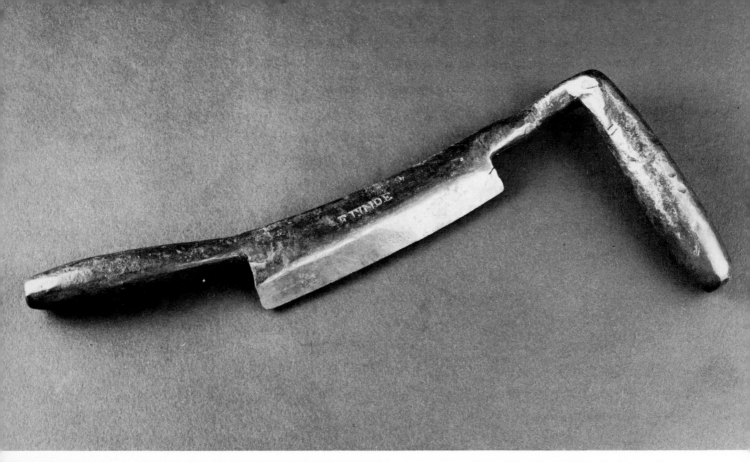

FIGURE 144. Chamfer knife. A form of heavy drawing knife, with one offset handle, which bevels the inner ends of staves. Length overall, 14 inches; blade, 6 inches.

FIGURE 145. Howel. Size of tool and curvature of the planing sole needed depend on type of cooperage being made. This massive example, with cutting iron almost three inches wide, was for large-capacity hogsheads. Length, 14⅛ inches; width, 6 inches.

FIGURE 146. Burn auger. The cooper heated it and used it to char bungholes to prevent rotting of the wood. Length, 20 inches; widest diameter, 1⅝ inches.

FIGURE 147. Cooper's shave. Finishing tool with sole that is slightly concave from toe to heel, for removing irregularities in the staves of completed cooperage. Not a common tool in America, this example was imported from Scotland. Width, 10 inches.

setting would measure six equal chords of the circumference. He then had the correct setting to scribe the diameter of the head. He sawed off excess material to bring the head to approximately the finished size and then worked it to exact dimension with a drawing knife. On the top and bottom edges of the circumference he shaped a bevel.

To prepare the barrel for inserting the heads, the cooper first leveled the inevitable slight variations in stave lengths with a sun plane (see Plate 25). This tool was shaped in a slight arc so that its surface could readily follow the curvature of the ends of the staves. To facilitate pressing the head into place, he cut a bevel with a chamfer knife on the inner edge of the staves, shaping the chime — the rim formed by the projecting ends of the staves. Then he planed a slight depression and smoothed irregularities on the inner surface of each end of the barrel, just below the chime, using a howel — a form of plane with a gouge-shaped cutting iron. (Prior to the introduction of the howel he used an adz with a curved blade.) Finally, he used a croze to form, in the middle of the howel cut, a groove into which the head would be seated. The croze was a grooving plane with an adjustable semicircular fence to permit different settings of the distance between the groove and stave ends. Various forms of the croze were equipped with a V-shaped, "saw-tooth," or "lance" cutting iron.

All surfaces of a well-finished cask were smoothed. For this purpose the cooper used a variety of drawing knives, shaves, and scorps. The scorp had a blade similar to that of a drawing knife, forged in a curve to shave concave surfaces, and provided with one or two handles. A round bunghole for filling the barrel was cut through the side with a bung borer (see Plate 23), and a tap hole bored on one head for later insertion of a spigot to draw off the contents. Some coopers used a burning iron in bungholes, to char and smooth the inner surface. Charring hardened the exposed end grain of the wood and

FIGURE 148. Bung pick. The pointed poll was for rapid extraction of a bung. Length of head, 8½ inches; handle, 11⅜ inches.

made the bunghole less subject to absorption of moisture and thus to rotting. Seams at the joints of the boards in the head, any slight openings between the staves, and the seating of the head were "flagged" to prevent leakage. Flagging — a material consisting of pithy rush stems — was forced into the cracks. The cooper used a flagging iron — a levering tool — to spread the staves apart to insert flagging.

The slack-barrel cooper followed essentially the same operations and used the same tools in making dry cooperage. His staves were considerably more pliable and could be drawn together for hooping by using a trussing rope wound around the end of the barrel and tightened with a winch.

There were over forty-three thousand coopers recorded in the United States Census of 1850, the majority located in New York, Pennsylvania, Ohio, and Indiana. By 1880 their number had grown to over forty-nine thousand, and there were in addition some four thousand artisans specializing in making staves, heading, or shooks. The first commercial production of oil at Titusville, Pennsylvania, in 1859, and the rapid growth of producing wells in the decades immediately following brought a boom in the coopering industry. "The demand for casks was so great," Kenneth Kilby writes, "that old whiskey casks were used, and oil magnates were buying forests to ensure a supply of timber for their coopers, producing 42-gallon casks."[8] This was the high point of employment in the trade — the boom was short-lived, as metal drums rapidly displaced wooden cooperage. Moreover, machine-made cooperage was a well-established product by the last quarter of the century, and alternative means for storing and shipping industrial and agricultural products were developed. A trade that had been of vital importance for centuries went into a decline, and in America today it is almost nonexistent.

FIGURE 149. Measurement of cask head. Laying out six chords of the circumference with the dividers gives the cooper a setting for radius of the head.

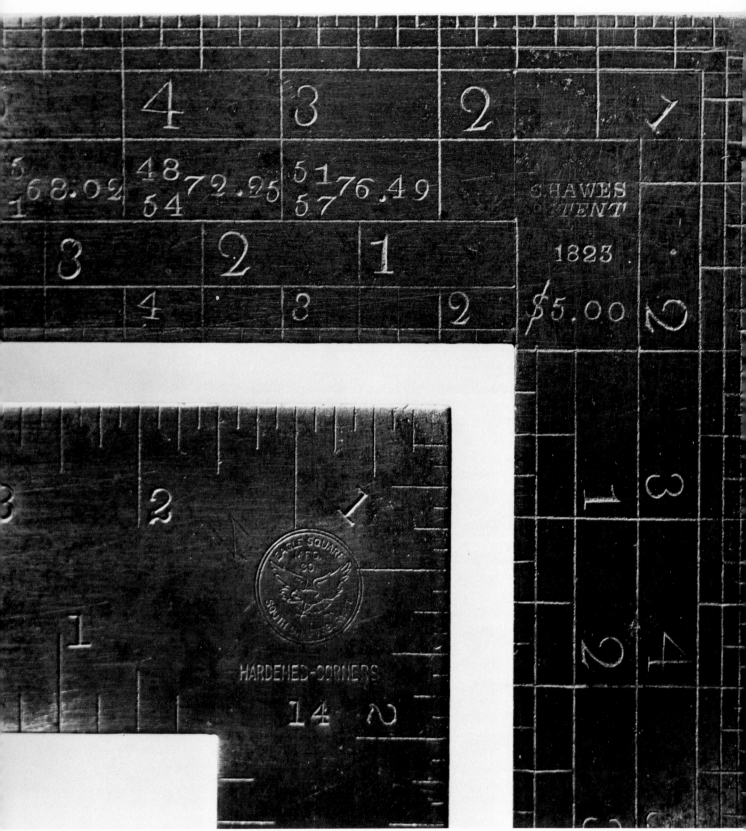

FIGURE 150. Steel carpenter's squares (details). *Above:* Hawes patent square of 1823 with brace rule on the blade; each marking is individually stamped. *Below:* Eagle square with inches and fractions pressed in the steel by Millington graduator machine. Blades of both are 24 inches long; length of narrower tongues varied but today is commonly 16 inches.

IX
TOOLS TO MEASURE
AND HOLD

STANDARDS FOR weights and measures had been the concern of royal authority from earliest times. In fourteenth-century England, Edward I decreed that three grains of barley, dry and round, laid end to end, represented one inch. This somewhat equivocal "standard" remained in effect for hundreds of years, and not until the nineteenth century were precise standards for yard, foot, and inch established.

Rule making as a specific trade probably started as early as the seventeenth century. Cask gauging, as we have seen, was at that time a recognized occupation, the gauger appointed by authority of a court or legislature. The rule was a common item among the tools of the carpenter, as contemporary accounts tell. Richard More, an exceedingly scholarly London carpenter, gives an interesting account of the status of the carpenter's rule in his book on the subject published in 1602.[1] In the second of two dedications (to "The Worshipful, the Master, Wardens, and Assistants of the Companie of Carpenters of the Citie of Londin: and to all others the curteous Readers"), More explained his reasons for writing: rulers then being made lacked standardization and the inch measurements varied; substantial errors of simple arithmetic were frequent because of one widely practiced — and faulty — method of computing board measure; the author hoped that as a result

of his explanations those who purchased timber would get fair measure. More recommended that his readers and his fellow members of the Carpenters' Company study geometry and learn how to use the rule, noting how he had profited from attending lectures on geometry at Gresham College. He explains and gives examples of the measurement of various shapes of timber, and tables for board and timber measurement.

More wrote at a time when mathematics and measurement were coming of age. Henry Briggs, who in 1596 was appointed the first Professor of Geometry at Gresham College, was shortly to work with John Napier on the invention of logarithms. It is likely that they were Briggs's lectures that the carpenter More attended. Edmund Gunter was then Professor of Astronomy at Gresham College. "Gunter's Line," a logarithmic notation that was applied to a slide rule, became a standard device for computation. It was included on many two-foot folding carpenter's rules as recently as the first years of the twentieth century. Richard More mentions having seen a new rule developed by Thomas Bedwell for determining board measure. In 1631 Bedwell's nephew, Wilhelm, published a description of this rule, telling how it differed from the usual carpenter's tool for computing board measurement.[2] His book in-

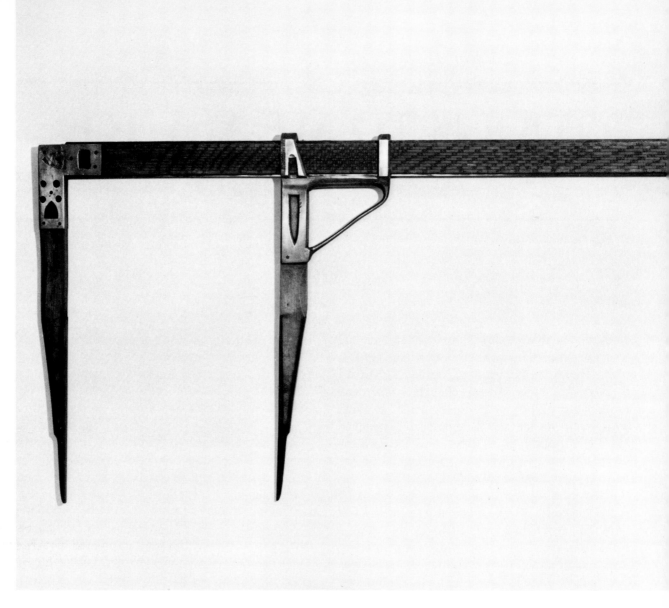

cludes a full-scale reproduction of the rule, and also a table "serving for the more exact, and speedy measuring of Board, Glasse, Stone, and such like, both Plaines and Solids."

Equitable weights and measures for business transactions were naturally important to the colonists. On March 8, 1637, within a year of the founding of the settlement at Hartford, the townsmen ordered that each plantation within the colony was to submit a measure to the next meeting of the court so that a consistent standard would be established. Eight years later, with additional towns added to the colony, there was a reassessment of the town weights and measures so that they could be "compared togather and made equall." A fine of twelve pence was imposed in 1647 for the sale of any commodity not measured by a rule or scale approved by the town clerk. He was required to be certain that rules were made of seasoned wood before affixing his seal and "to breake or demolishe such wayghts, yards or measures as are defective."[3]

Rod, rule, and scale making was apparently a well-established trade by the mid-eighteenth century. In *The New-York Mercury* of May 27, 1754, James Ham described himself as "mathematical instrument-maker, at the house

FIGURE 151. Lumber caliper. An itinerant log-rule maker, William Greenlief, fashioned this measuring device of the lumberman. The wheel measures the length of a log, five feet to each rotation. The caliper measures the diameter of the log in inches. The scale on the rule gives a figure for the quantity of wood that can be obtained from the felled tree. Calipers were scaled to read in cord measure of sawn logs, in cubic feet, or in board feet of sawn lumber. Length overall, 54½ inches.

wherein the Widow Ratsey lately lived, near the Old-Dutch-Church, in Smith's Street," making and selling a variety of instruments in wood, brass, and ivory, including gauging rods, sliding and Gunter's scales, and carpenters' bevels.[4] By the early 1800s rule making had become a factory business. Belcher Brothers were in operation at 146 Division Street, New York City, in 1824; Solomon A. Jones at Hartford, Connecticut, in 1838; and Joseph Watts in Charlestown, Massachusetts, and later in Boston, from 1834.

In 1838, Edward A. Stearns of Brattleboro, Vermont, bought the rule factory started five years earlier by S. Morton Clark. Stearns built a thriving business, which he operated until his death in 1856. The company was later purchased by the Stanley Rule and Level Company, and the rule-making machinery moved to New Britain about 1870. The reputation of Stearns boxwood and ivory rules was such that Stanley continued to use the name on some rules of its manufacture for another thirty years. By the end of the century, Stanley had captured the major share of the American rule market.

In the *Mechanick Exercises*, Moxon described and illustrated the one-foot rule, the ten-foot rod for use in house building, the bevel,

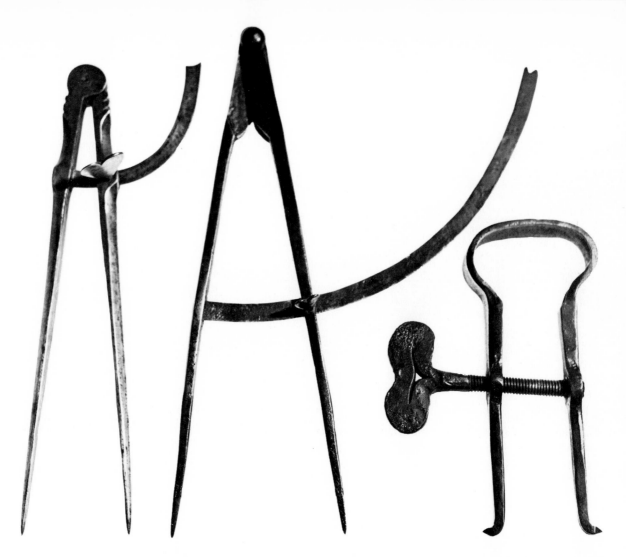

square, miter square, marking gauge, plumb, and level. To determine true horizontal, for work such as setting sills and joists of a house, the carpenter used the level. In an early form, the carpenter's level consisted of a triangular wooden frame with a board from two to ten feet long at the base. From the apex of the triangle, a lead weight was suspended by a string. When the string was perpendicular to the board at the base, coinciding with a reference point or line, the base was level, or horizontal. Plumb, or a true vertical, was established by the plumb bob weighting a suspended string. The carpenter's spirit level was an invention of the late seventeenth century. A slightly arched, sealed glass tube filled with alcohol or other liquid (leaving a small bubble of air) is set into the wooden body of the level. When the bubble is centered in the tube, the bottom surface of the tool is level. A second tube, set toward one end of the tool and at right angles to its length, establishes a vertical or plumb reading when that bubble is centered. Spirit levels were made by some instrument-makers in the early nineteenth century, such as H. M. Pool of Easton, Massachusetts, and later by many manufacturers of carpenters' and joiners' tools. Among the latter, the Stanley firm, the Davis Level and Tool Company of Springfield, Massachusetts, and the Chapin-Stephens Company of Pine Meadow, Connecticut, were prominent. In heavy mahogany, rosewood, or cherry, with brass end plates or fully brass bound as in the more expensive models, the level was a serviceable and handsome tool.

The marking gauge and mortise gauge were carpenters' and joiners' tools for scribing construction lines in laying out work (see Plate

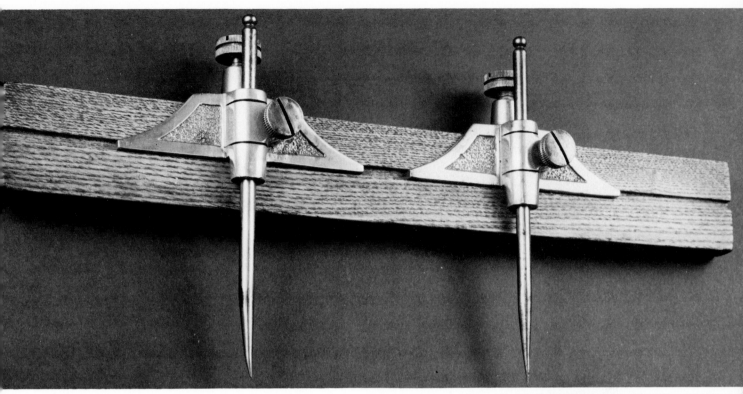

FIGURE 152. (*Opposite.*) Shapely forging in iron. Dividers for measuring and scribing arcs and circles. *Right:* Inside caliper wrought from a rasp originally made to trim horses' hooves.

FIGURE 153. (*Above.*) Trammel points. They form a beam compass to lay out larger circular shapes in wood. Cast-brass trammel heads with steel points. Length of points, 7 inches.

FIGURE 154. (*Right.*) Double caliper. The legs measured the inner diameter, the upper part the outer. Found in Pennsylvania. Length: 9 inches.

FIGURE 155. Blacksmith-made squares (details). *Left:* Dated 1824, it is typical of the hand-forged tool. Blade and tongue are hammer-welded at the joint. *Below:* Hand forged with hand-struck numerals and graduations.

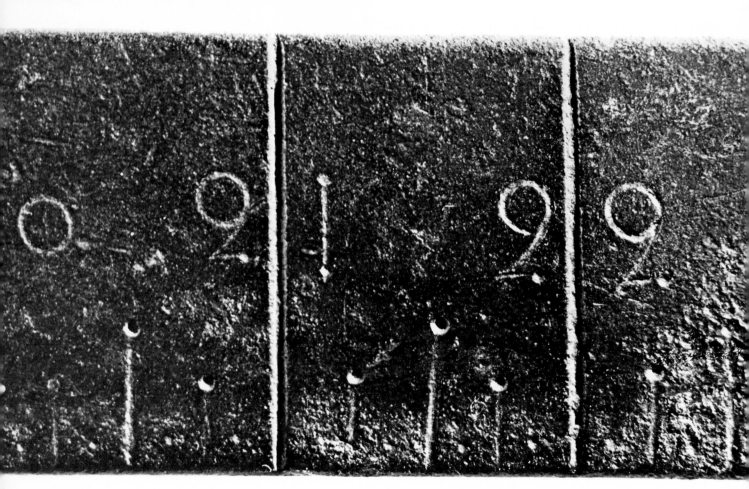

31). A scribed line served as a guide in sawing, planing, outlining a mortise, or otherwise working wood to specific dimensions. The marking gauge had one scribing point, and the mortise gauge a second, adjustable point; both frequently had a scale of inches on the shaft, to adjust the movable head to a specific distance from the scribe points.

Woodworkers used the dividers, or compass, to determine proportional measurements, to scribe circles and arcs, or to transfer a measurement from one piece of work to another. For laying out large circles and ellipses, the beam compass, or trammel points, was used. This tool had two adjustable points which could be moved to any positions on the wooden beam. Bevels were also constantly used; the tongue of the adjustable bevel was set to a prescribed angle and used as a straightedge for marking angles.

Squares were indispensable, especially for laying out construction lines and checking right-angle joints. The joiner and cabinetmaker normally used small try squares with blades three to eighteen inches long, which they frequently fashioned themselves from choice pieces of mahogany. The manufactured joiner's square, with a steel blade and rosewood or ebony handle, brass trimmed, was appropriate for the craftsman who handled fine tools with care; for rough and ready use, it was also manufactured wholly in metal. House- and barn-builders' squares made from wood or iron often reached considerable size, with blades from six to eight feet long.

The usual carpenter's square was blacksmith made in iron into the nineteenth century, or factory made in steel from 1820. The latter, with two-foot blade and fourteen- to eighteen-inch tongue, has many functions in building apart from its basic uses — marking right angles and measuring feet and inches. Board

FIGURE 156. Davis levels. Adjustable spirit level, plumb, and inclinometer are combined in these iron tools patented in 1867, made for carpenters and machinists. Lengths, 6 inches and 12 inches.

FIGURE 157.   Bench dogs. Used to hold a workpiece temporarily to the bench top.
Widest dimensions: left, 4½ inches; right, 6 inches.

measure can be calculated by the scale on the back of the blade. A version of this scale, known as the "Essex board measure," was developed by Jeremiah Essex, a nineteenth-century square-maker of North Bennington, Vermont. Another series of numbers constituting the "brace rule" is struck on the middle of the tongue. Various dimensions for two sides of a right triangle are given, and to the right of each pair of figures the length of the hypotenuse of that triangle is shown. This scale was used to measure lumber for diagonal braces in house, barn, or mill construction. Other scales were sometimes added to the square; one provided the measurement for the length of roof rafters at various pitches of the roof.

Local Vermont tradition credits Silas Hawes of South Shaftsbury with the "invention" of the steel square in the early nineteenth century. He is said to have taken some old saw blades in payment for shoeing the horse of an itinerant peddler, and welded two of these together to make the first steel square. The story has

charm, but is apocryphal; steel squares were known at least two hundred years earlier. Among the tools that belonged to Francis Eaton, the carpenter of Plymouth Plantation, was an iron square. When another Eaton, Governor Theophilus Eaton of New Haven, died in 1657, two iron squares were among his many tools. Silas Hawes did, however, receive a patent on the carpenter's square from the United States Patent Office on December 15, 1819. Unfortunately, because most of the patent records of the early nineteenth century were lost in fires, the features Hawes patented will probably never be known. Early examples of Hawes squares, dated 1823 and 1826, include a board measure, and the flat blade and tongue were tapered. Tapering improved the "hang" of the tool by making it more manageable in the carpenter's hands. Possibly one or both of these details were improvements on which the inventor was granted the patent.

South Shaftsbury, where Hawes started his business, became the country's steel-square

188

manufacturing center during the nineteenth century. Several local manufacturers made squares on the Hawes patent. Between January 1 and April 1, 1859, square-makers Dennis George, Jeremiah Essex, Herman Whipple, and the firm of Hawks, Loomis & Company joined forces to form the Eagle Square Company. The success of this new company was in no small measure attributable to their control of the "Millington graduator," a machine patented by Norman Millington and Dennis George on August 8, 1854. Their device mechanically stamped the graduations of the inch on the tools. Prior to that time, every numeral, point, ruled line, and fractional demarcation of the inch was hand struck with dies and hammer; there were up to eight hundred fractional graduations on one side of the rule alone. The original design of the machine marked only eighths of an inch, but it was soon improved to mark in sixteenths. Following the death of some of the founders and sales of stock, the company changed hands. It was incorporated in 1874 under a new management, capitalized at $60,000. From that date, with some lapses in lean years, the firm grew progressively and paid its owners an excellent return. In 1916 the Stanley Rule and Level Company purchased the company, continuing the Eagle name and the South Shaftsbury location. Today the Eagle Square Manufacturing Company, producing Stanley squares and other proprietary brands, has eighty-five percent of the world market for steel carpenter's squares, and produces one and a half million squares annually.

A work surface and work-holding devices have always been essential for the craftsman using hand tools. The carpenter and shipwright worked much of the time out of doors. Their sawing, hewing, fitting, and joinery were performed while the work was set on sawhorses, trestles, or on temporary staging. The wheelwright had special devices for holding felloes while he shaped them, for gripping the wheel

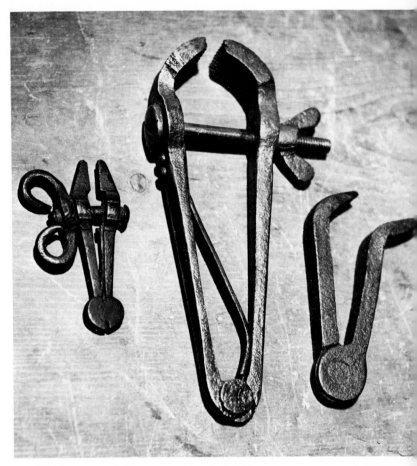

FIGURE 158. Hand vises. Handy gripping tools for working with small parts. Vise at right is associated with metalworking, often with the locksmith. Lengths, left to right: 4¼, 9⅝, and 5½ inches.

hub as he bored and shaped the mortises for spokes, and for holding the wheel while he fitted the iron tire. There were also many general types of holding devices, of which several were directly associated with the working surface of the artisan's bench.

The workbench, used to support and position wood while it is shaped, was used in many trades. It was essential for the joiner and cabinetmaker, practical for general carpentry, and it has been used in some form for thousands of years. The Roman joiner used a low, narrow

FIGURE 159. Portable table vise. Used by cabinetmaker or joiner to hold small metal fittings. Length, exclusive of clamp, 5¼ inches; width of jaws, 2⅛ inches.

190

FIGURE 160.   Hand screws. Common woodworker's clamp to hold material, as in glueing. Overall length of jaws, 10 inches.

bench with four legs, about eighteen inches from the ground. He sat astride one end with his work in front of him. This form of low worktable continued in use through the Middle Ages to recent times. Various clamping devices and pegs were attached to the side or run through the surface to hold the workpiece. Ants Viires, in tracing the development of benches used by craftsmen in Estonia, has described several versions of the sit-upon style used in that country, noting that types still in use by coopers in one district of Estonia are identical to those used by medieval cabinetmakers of Nuremberg.[5] The typical American cobbler's bench was basically of this construction, and a similar type was used by the makers of handmade nails. The American cooper used a low bench known as a "shaving horse." A lever passed through an aperture in the top of this bench and was pivoted at that point; the top end of the lever was shaped to grip a barrel stave as the cooper pressed his feet against a cross bar or pedal on the bottom end. The carpenter sat at the same sort of bench to taper pine and cedar shingles with his drawing knife.

It was the invention of the plane, Goodman suggests, that made the table-high bench necessary. A short piece of wood could be planed on a low bench, but to plane a long piece on its entire length the artisan needed to walk along it. A wooden bench vise with two gripping surfaces, brought together by tightening one or two large wooden or iron bench screws, supported the wood or one end of it. Several rectangular holes, spaced vertically on the legs of the bench, accepted square pegs on which one end of a long board could be rested while the vise gripped the other end. Control of shorter pieces was also possible with the use of the bench stop, an iron pin with toothed edge set at a right angle to its vertical shaft. One end of this tool was set into a hole in the bench top, and the workpiece was forced against the toothed edge of the stop. The joiner used still other methods of holding

material on the bench. In one method, two sharp points of a tool known as a bench dog were driven into the bench top and the third point set into the workpiece; two or more dogs could be used to fasten the work.

The portable table vise, which could be clamped to the edge of bench or worktable, gripped small pieces of wood or metal. This tool was often blacksmith made, rather than a factory product. The hand vise, although not essentially a woodworking tool, was also found with woodworking tools because it was useful for providing a firm handhold for small pieces of material that required filing or shaping.

The ship carpenter, the joiner, and many other artisans made frequent use of clamps, both big and small, short and long, of wood or of iron, for temporarily holding material together. The cabinetmaker gluing joints, and the joiner gluing together parts of doors or windows, needed several clamps on one piece of work. The shipwright assembling the frames of a ship clamped the futtocks together with large iron C-clamps. The overlapping futtock sections were bored with an auger and pinned with treenails, after which the clamps were removed. Long, adjustable bar clamps and the smaller hand screws were typical clamping devices. The gripping action of many holding devices was achieved by means of the screw. The screws of bench vises, hand-screw clamps, and bar clamps were customarily made of wood, rather than iron, up to the late nineteenth century. The threads of the screw were cut with a screw box, a wooden or metal stock containing one or more V-shaped cutting blades, which was rotated around the stem. The matching thread in which the screw rotated was cut by a screw tap into the walls of a hole bored in a second piece of wood. This tap was forged from iron with a series of lateral cutters, or, later, with one V-shaped cutter. The cutters of the tap were designed to function in an inclined plane corresponding to that of the screw box cutter.

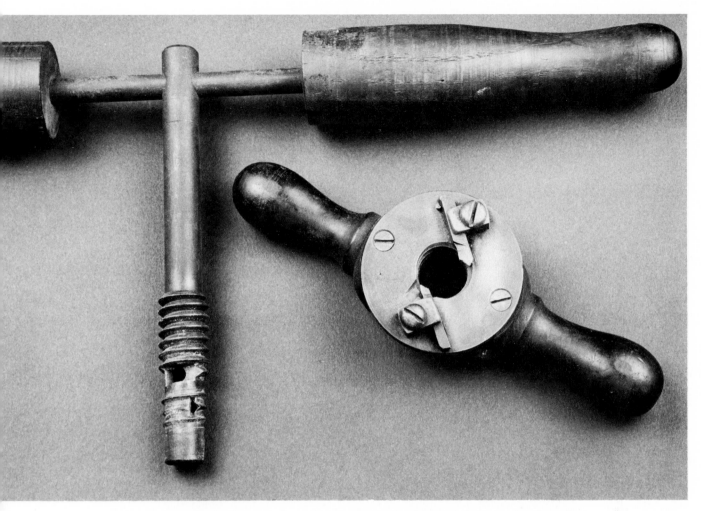

FIGURE 161. Tap and screw box. Hole is threaded with the 1-inch tap (*left*), and matching screw is threaded with the screw box (*right*). Full length of T-handle, 20 inches; box 9⅛ inches.

FIGURE 162. The ultimate combination plane. Stanley's "Patent Universal '55' Plane." Marketed first in 1897, it was discontinued in 1962. Length, 11½ inches.

X

DECLINE OF
THE HAND TOOL

AT THE close of the eighteenth century, conditions in the United States favored major changes in industrial life. Many of the conditions required for introducing machinery in manufacturing processes could now be met: a growing home population and foreign trade offered larger markets. Basic raw materials, including wood and iron, were plentiful. Capital available for investment in machinery had increased in the hands of the merchant class through the latter part of the 1700s. Transportation had improved; water transport was inexpensive, and if highways were minimal and difficult to travel, the need to extend and improve them was at least now recognized. The development of industrial machinery in England and in Europe demonstrated its potential. Finally, inventiveness was being promoted and protected by rights of patent.

Machinery began to replace hand labor very early in the nineteenth century. Woodworking machines in particular had a far-reaching impact. In 1795 Oliver Evans of Philadelphia wrote a book for millwrights and millers in which he described a mill that was, in effect, fully automated, from depositing the grain to bagging the flour. He included an illustration of an improved water-powered sawmill that advanced the wood mechanically as it sawed the length of a log, stopped cutting when the log

had been sawed to within three inches of its end, and returned the log on a sliding carriage, now ready for sawing the next plank. "One mill attended by one man, if in good order, will saw more than 20 men with whip-saws, and much more exactly."[1] Evans was not romancing; such mills were built. In fact, machines had been replacing sawyers for well over a century. There were numerous water-powered sawmills throughout the country; their up-and-down blades were soon to be replaced by faster-cutting circular saws powered by steam. At Brunswick, Maine, in 1820, a large rotary sawmill was in operation, cutting large timbers and manufacturing clapboards, barrel staves, and heading. In the same year, Thomas Blanchard of Middlebury, Massachusetts, eliminated hand shaping of gunstocks with a patented shaping lathe for turning irregular forms. Where there was no source of power at hand, a variety of horse-powered machines could be used to saw timber and shingles. By midcentury a steam-powered portable sawmill that could be set up in the forest was on the market.

Machinery for finishing lumber also came into use in the early nineteenth century. Samuel Bentham of England is considered the inventor of the first planing machine. This 1791 device functioned with fixed cutters that dressed wood held on a reciprocating bed. Bentham and

195

FIGURE 163. Compass plane in wood. For planing a slightly concave surface. Length, 8 inches.

others next developed the planer with cutters held in a rotating cylinder — the type of planing machine manufactured today. The planing machine, and the grooving, routing, and molding machines that rapidly followed, ultimately made hand planing a thing of the past. In 1828 William Woodworth of Hudson, New York, received a patent on a machine for planing, tonguing, grooving, and cutting boards. J. Leander Bishop states: "This patent is remarkable for the amount of litigation arising out of it for many years after, and for having been longer extended than any other patent, as well as the great profits it has yielded to the owners."[2] Knight, in his *American Mechanical Dictionary*, complains that the Woodworth planer "became an odious monopoly, and did much to discredit the patent system" — the observation is amusing, coming from the editor of the *Official Gazette* of the United States Patent Office. By about 1850, Woodworth and subsequent owners of the patent had licensed one thousand machines throughout the country. They maintained control of a large share of the market for

planing machines by attractive leasing arrangements (which stipulated a fixed minimum retail price per planed board foot) and by numerous suits against manufacturers of competitive machines. In argument before the Senate and House Committees on Patents, Woodworth's assignees attempted unsuccessfully to obtain a third extension of the original patent to 1870. Opponents stated that "the effect of the monopoly is to keep the price of planing by machine just on the verge of the cost of doing the same work by hand."[3]

Planing and molding machinery that could easily produce ten thousand feet of finished work per day affected the woodworking trades in many ways. Finished boards, flooring, molded cornices, baseboards, picture molding, banister rails, door and window architraves, doors, and window sash became stock items of a mill or lumberyard, rather than being made on the job by the artisan. Desk, bedstead, dresser, and other furniture once made by the cabinet-maker became factory products of the saw and planing machine. The impact on style in joinery

FIGURE 164.   Cast-iron and steel version of the compass plane. Length, 10¼ inches.

and cabinetmaking was great. Moldings became flatter, for example, because flatter moldings could be produced faster and used less wood.

The carpenter and builder were also profoundly affected by another industrial innovation — structural iron, used for the iron bridge and iron-framed building. The first American iron bridge spanned the Hudson River at Waterford, New York, in 1804. Structural cast-iron framing for large buildings began to replace wood timbers about 1850; ten years later cast- and wrought-iron girders were commonplace.

These innovations of machine and material led to the virtual elimination of the traditional hand woodworking trades. In 1882 Henry Hall wrote: "For general trading, especially for long voyages, there is every reason to believe that sailing ships will continue to be built, as they can carry a cargo at a lower rate of freight than steamers." Comparing the cost per ton of ships constructed of iron with those of wood, he saw little prospective demand for iron sailing ships.[4]

In this assumption he was correct: few iron sailing ships were to be built. But he was rapidly proven wrong in thinking that trade would continue to move in wooden bottoms. In the 1850s, when the merchant sailing fleet was at its peak, it was already technologically obsolete. The century closed with American shipping in a decline, having lost out to the iron and steel merchant steamers of Britain.

Carriage-making as a trade was to last into the first quarter of the twentieth century, but the manufacture of vehicles and wheels had long since been reduced from a skilled craft to the assembly of manufactured parts.

New materials, refinements, and diversification in uses were applied to hand tools in what were to be vain efforts to compete with the machine. The evolution of the plow plane during the thirty years from 1867 to 1897 typifies these efforts. The traditional plow plane, made of wood with iron and brass fittings, was altered structurally in 1867. In that year Russell Phillips of Gardiner, Maine, patented a model of the tool made from cast iron. The patent

FIGURE 165.   Iron smooth plane. A Derby, Connecticut, firm made a series of bench planes of this pattern in cast iron. The winged design at center of the casting is an unusual decorative feature. Length, 9 inches.

covered relatively inconsequential changes in the shape of cutting irons and in the means for adjusting the depth of the groove cut. Its significance was in the transition of the tool from a product of the plane-maker working in wood to one of a foundryman casting in iron.

Diversification took the form of the combination tool. In 1870, Charles Miller, of Brattleboro, Vermont, was granted a patent on another cast-iron plow plane; this tool could cut matching tongue-and-groove boards and be converted to a tool for shaping rabbets by attaching an iron fillister bed to the stock. Miller was fortunate; his patent was shortly taken up by the Stanley Rule and Level Company, and Miller moved to New Britain to work. This particular plane was in production and advertised by Stanley within six months, offered in either iron, at fifteen dollars, or in gunmetal at eighteen dollars. A wood plow plane, in ebony, boxwood, or rosewood, with ivory-tipped adjusting arms, sold at that time for ten dollars, and a more modest plow plane in beech for four to six. In spite of the high price, Miller's tool was success-

ful because it was practical, well made, and widely marketed. It became the progenitor of a line of Stanley planes of which the "Stanley 45" and the "Patent Universal '55' Plane" were the lineal descendants. The latter, introduced in 1897 and advertised as "A Planing Mill Within Itself," was the ultimate in the combination plane. In the hand of a practiced carpenter it could plow grooves, shape dadoes and rabbets, and form a limitless number of moldings with its accompanying four boxes of fifty-two or more special cutting irons. Several manufacturers tried, with only marginal success, to compete with Stanley in producing combination planes. Those made by Otis Smith in Rock Fall, Connecticut, and Jacob Siegley in New York were but two of the competitive tools made of cast iron. It is an anomaly that these tools, popular between 1870 and the early 1900s, had a sizable sale at a time when mill-produced lumber had otherwise captured the market in shaped wood. The planing mill had also preempted flat planing of wood, and the bench plane had become a finishing and fitting tool

198

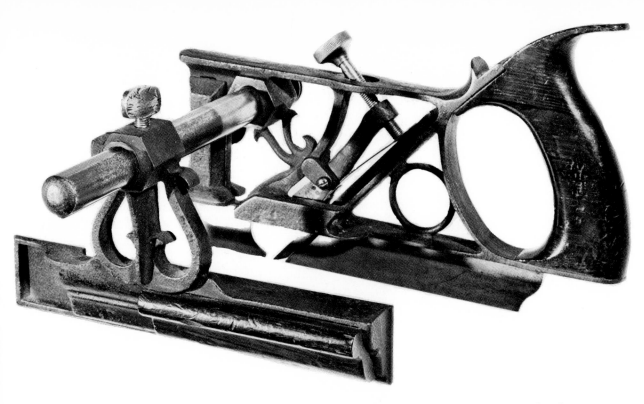

FIGURE 166. The foundryman/plane-maker. Cast iron takes the place of beech, boxwood, and rosewood in the grooving plow plane. Russell Phillips's patent plane of 1867. Length, 10⅝ inches.

rather than one for surfacing rough-sawn wood.

Goodman has plotted a curve of the number of known and dated plane-makers working in the British Isles from 1700 to 1960. This curve peaked at about 1855. No such comprehensive study has yet been done for American plane-makers. Lists and working dates have been provided for Boston plane-manufacturers by William Hilton,[5] for western Massachusetts makers by Elliot Sayward and William Streeter,[6] and a great number of manufacturers in New York and several other states are listed by Kenneth Roberts.[7] These lists indicate that the number of manufacturers in these areas reached a high point in the thirty years between 1830 and 1859. After 1879 the number fell off drastically, not only because of technological change but because manufacturing became concentrated in a limited number of companies.

The tools of the cooper met the same fate as the planes of the carpenter and joiner. Knight's *Dictionary* of 1876 was replete with descriptions of crozing, hoop-bending, hoop-riving, hollowing and backing, stave-cutting, stave-saw-ing, and stave-jointing machinery. In the period from 1850 to 1890, when the population almost tripled to sixty-three million, the number of coopers increased only from forty-four thousand to fifty thousand. Of the fifty thousand, the majority were machine operators in barrel factories, rather than "coopers." In March 1907 the pages of the *National Coopers' Journal* were filled with advertisements for the "Rochester Slack Barrel Stave Jointer," the "Automatic Triple Hoop Planer," and similar machinery. An article described a new chamfering, crozing, and howeling machine, "Capacity, 1,000 to 2,000 barrels per day. Boxed for export, 3,800 pounds." The Hynson Tool & Supply Company of Saint Louis was waging a losing battle; it was the only firm still advertising coopers' hand tools. It displayed but three, and noted that one of them, its bung borer, was "made for both hand and machine." Within forty years even the machines of the cooperage industry would be obsolete.

The market for hand woodworking tools exists today only to meet the needs of the home handyman, the few men and women in cabinet-making and woodworking crafts who prefer and can afford to work by hand, and carpenters and others in the building trades. The latter still use a limited number of hand tools because of their convenience, portability, suitability to a task, or because no machine has yet taken their place.

Those interested in the collecting, preservation, or in the occasional use of the woodworking tools of the eighteenth and nineteenth centuries find that the demand for antique tools is growing and the supply diminishing. The era when hand craftsmanship was the rule has long passed; the tools of the trades that built our nation have been for the most part treated as curiosities, or worse, gone unrecognized, always in danger of being relegated to undeserved oblivion. Now their purpose, their significance in our history, and their honest character and beauty of form are at last being acknowledged and valued.

FIGURE 167. Spill plane and spills. The plane's sharply skewed blade forms a tightly curled shaving for transferring fire from hearth to candles or a pipe. Once a common household implement, spill planes met an everyday need. Today, they are rare and sought as example of a bygone life-style. Length, 10⅞ inches.

REFERENCES

I / THE CULTURAL HERITAGE

1. Ivor Noël Hume, *Historical Archaeology* (New York: Alfred A. Knopf, 1976), p. 7.
2. W. L. Goodman, *British Plane Makers from 1700* (London: G. Bell & Sons, 1968), pp. 13, 16.
3. Samuel McKee, *Labor in Colonial New York, 1664–1776* (1935; reprint ed., Port Washington, N.Y.: Ira J. Friedman, [1965?]), pp. 83–84, 88.
4. Exhibition of the Works of Industry of All Nations, London, 1851, *Reports by the Juries* (London: W. Clowes, 1852), p. 201.
5. Nathaniel B. Shurtleff, ed., *Records of the Governor and Company of the Massachusetts Bay in New England*, vol. 2, 1642–1649 (Boston: William White, 1853), p. 401.
6. U.S. Bureau of the Census, *Historical Statistics of the United States, Colonial Times to 1970*, vol. 2 (Washington, D.C., 1975), p. 294.
7. Tench Coxe, *A View of the United States of America, in a Series of Papers Written at Various Times Between the Years 1787 and 1794* (Philadelphia: William Hall, 1794), p. 420.
8. U.S. Bureau of the Census, *Historical Statistics*, vol. 1, p. 541.
9. U.S. Patent Office, *Subject-Matter Index of Patents for Inventions Issued by the United States Patent Office from 1790–1873, Inclusive*, 3 vols. (Washington, D.C.: Government Printing Office, 1874).

II / HISTORICAL OVERVIEW

1. Kenneth P. Oakley, *Man the Tool-Maker* (Chicago: University of Chicago Press, 1957), p. 23.
2. S. A. Semenov, *Prehistoric Technology* (London: Cory, Adams & Mackay, 1964).
3. Vladimír Kozák, "Stone Age Revisited," *Natural History* 81, no. 8 (1972): 14–24.

4. Oakley, *Man the Tool-Maker*, p. 36.
5. Leslie Aitchison, *A History of Metals*, vol. 1 (New York: Interscience Publishers, 1960), p. 26.
6. P. L. Shinnie, *Meroë; a Civilization of the Sudan* (New York: Frederick A. Praeger, 1967), pp. 160–161.
7. W. M. Flinders Petrie, "History in Tools," Smithsonian Institution, *Annual Report*, 1918, pp. 563–572.
8. W. M. Flinders Petrie, *Tools and Weapons* (London: British School of Archaeology in Egypt, 1917), p. 19.
9. Josef M. Greber, *Die Geschichte des Hobels* (Zurich: VSSM-Verlag, 1956), pp. 54–57.
10. Petrie, *Tools and Weapons*, p. 39.
11. W. L. Goodman, *The History of Woodworking Tools* (London: G. Bell and Sons, 1966), p. 43.
12. Greber, *Die Geschichte des Hobels*, pp. 98–99.
13. Petrie, "History in Tools," pp. 565–566.

III / SOME BASIC TOOLS

1. J. P. Brissot de Warville, *New Travels in the United States of America. Performed in 1788* (London: J. S. Jordan, 1792), p. 127.
2. Nathaniel B. Shurtleff, ed., *Records of the Governor and Company of the Massachusetts Bay in New England*, vol. 2, 1642–1649 (Boston: William White, 1853), pp. 61–62.
3. Cited in John W. Oliver, *History of American Technology* (New York: Ronald Press Co., 1956), p. 70.
4. James Thatcher, "Observations upon the Natural Production of Iron Ores . . . in the County of Plymouth," in *Collections of the Massachusetts Historical Society*, vol. 9 (Boston, 1804), p. 264.
5. Raphael Salaman, "Tradesmen's Tools c1500–1850," in *A History of Technology*, ed. Charles Singer [et al.], vol. 3 (Oxford: Clarendon Press, 1957), p. 110.

6. François A. F. duc de La Rochefoucauld-Liancourt, *Travels Through the United States of North America . . . 1795, 1796, and 1797*, vol. 1 (London: R. Phillips, 1799), p. 165.
7. William Douglass, *A Summary, Historical and Political . . . of the British Settlements in North-America*, 2nd ed., vol. 2 (London: R. and J. Dodsley, 1760), p. 52.
8. J. Leander Bishop, *A History of American Manufactures from 1608 to 1860*, 3rd ed., vol. 3 (1868; reprint ed., New York: Johnson Reprint Corp., 1967), p. 213.
9. The tools here noted were included in lists prepared for a meeting of the Early American Industries Association at Plymouth, Massachusetts, in May 1970. Cited hereafter as *Plymouth Inventories.*
10. *Public Records of the Colony of Connecticut, Prior to the Union with the New Haven Colony, May, 1665* (Hartford: Brown & Parsons, 1850), pp. 463–464.
11. Bishop, *History*, vol. 3, p. 300.
12. John S. Kebabian, "The Disston Factory, Philadelphia, Pennsylvania," *Chronicle of the Early American Industries Association* 23 (1970): 39–40; Henry Disston & Sons [*Catalogue*] (1914; reprint ed., Lancaster, Mass.: R. K. Smith, 1976), pp. 2, 124.
13. W. M. Flinders Petrie, *Tools and Weapons* (London: British School of Archaeology in Egypt, 1917), p. 39.
14. Robert H. Carlson and T. A. Stevens, "On the Origin of the Spiral Auger," *Chronicle of the Early American Industries Association* 20 (1967): 53.
15. U.S. Patent Office, *Report of the Commissioner of Patents*, 1859, 1:666.
16. U.S. Patent Office, *Report*, 1864, 1:507.
17. *American State Papers. Documents, Legislative and Executive, of the Congress of the United States . . . Commencing March 3, 1789, and Ending March 3, 1815*, vol. 5: *Finance*, vol. 1 (Washington, D.C.: Gales and Seaton, 1832) pp. [5]–8.
18. *American State Papers*, vol. 5: *Finance*, vol. 1, p. 139.
19. *Iron Age* 12 (1873): Oct. 16, p. 13.
20. Charles F. Hummel, "English Tools in America: the Evidence of the Dominys," *Winterthur Portfolio* 2 (1965): 28.
21. *Pennsylvania Packet, and Daily Advertiser*, no. 2943, July 9, 1788.
22. U.S. Patent Office, *Report*, 1858, 2:345.
23. U.S. Patent Office, *Report*, 1867, 2:1035.
24. Stanley Rule and Level Company, complainant, *Copy of Evidence, United States Circuit Court, District of Connecticut, In Equity: the Stanley Rule and Level Company, Against Leonard Bailey* (Hartford: Stone & Mills, 1878), pp. 90–91.
25. *Public Records of the Colony of Connecticut . . . May, 1665*, p. 473.

IV / CARPENTER AND SAWYER

1. John Winthrop, *Papers*, vol. 3, 1631–1637 (Boston: Massachusetts Historical Society, 1943), p. 21.
2. William Wood, *New Englands Prospect* (London: John Bellamie, 1634), p. 53.
3. Winthrop, *Papers*, vol. 3, pp. 4, 42, 155.
4. Winthrop, *Papers*, vol. 3, pp. 161–162, 208.
5. *Plymouth Inventories.*
6. Winthrop, *Papers*, vol. 4, pp. 11–12.
7. Norman M. Isham and Albert F. Brown, *Early Connecticut Houses* (1900; reprint ed., New York: Dover Publications, Inc., 1965), p. 289.
8. Nathaniel B. Shurtleff, ed., *Records of the Governor and Company of the Massachusetts Bay in New England*, vol. 1, 1628–1641 (Boston: William White, 1853), pp. 74, 76, 109, 159, 183, 326.
9. *Public Records of the Colony of Connecticut, Prior to the Union with the New Haven Colony, May, 1665* (Hartford: Brown & Parsons, 1850), p. 65.
10. Charles E. Peterson, "Early Lumbering: A Pictorial Essay," in *America's Wooden Age: Aspects of its Early Technology*, ed. Brooke Hindle (Tarrytown, N.Y.: Sleepy Hollow Restorations, 1975), p. 66.

V / JOINER AND CABINETMAKER

1. W. L. Goodman, "Tools and Equipment of the Early Settlers in the New World," *Chronicle of the Early American Industries Association* 29 (1976): 48–49.
2. Daniel M. Semel, "A First Look at Duncan Phyfe's Tool Chest," *Chronicle of the Early American Industries Association* 29 (1976): 58.
3. John Quinan, "Asher Benjamin as an Architect in Windsor, Vermont," *Vermont History* 42 (1974): 188.
4. Henry C. Mercer, "The Dating of Old Houses" (paper read at a Meeting of the Bucks County Historical Society, New Hope, Pa., October 13, 1923), pp. 14, 16.
5. House Carpenters and Joiners of the City of Cincinnati, *Book of Prices, Adopted Monday, January 4, 1819*, rev. and enl., February, 1844, by Louis Shally (Cincinnati: L'Hommedieu & Co., 1844), pp. 20–21.
6. Joseph K. Ott, "Rhode Island Furniture Exports 1783–1800," *Rhode Island History* 36 (1977): 5.

VI / SHIPWRIGHT

1. William Bradford, *Of Plymouth Plantation, 1620–1647*, ed. Samuel Eliot Morison (New York: Alfred A. Knopf, 1952), pp. 146, 155.
2. Nathaniel B. Shurtleff, ed., *Records of the Governor and Company of the Massachusetts Bay in New England*, vol. 1, 1628–1641 (Boston: William White, 1853), pp. 394, 402.
3. Shurtleff, *Records of Massachusetts Bay*, vol. 1, pp. 337–338.

4. Henry Hall, "Report on the Ship-Building Industry of the United States," pp. 50–59, in *Tenth Census of the United States*, vol. 8, *Special Reports* (Washington, D.C., 1884).

5. *Public Records of the Colony of Connecticut, May, 1678–June, 1689* (Hartford: Case, Lockwood, 1859), p. 299.

6. *American State Papers. Documents, Legislative and Executive, of the Congress of the United States . . . Commencing March 3, 1789, and Ending March 3, 1815*, vol. 5: *Finance*, vol. 1 (Washington, D.C.: Gales and Seaton, 1832), p. 140.

7. *Public Records of the Colony of Connecticut, Prior to the Union with the New Haven Colony, May, 1665* (Hartford: Brown & Parsons, 1850), pp. 473–474.

VII / WHEELWRIGHT AND CARRIAGE-MAKER

1. Joseph Needham, *Science and Civilization in China*, vol. 4, pt. 2: *Mechanical Engineering* (Cambridge: University Press, 1965), p. 13.

2. Nathan Perkins, *A Narrative of a Tour Through the State of Vermont . . . 1789* (Woodstock, Vt.: Elm Tree Press, 1920), p. 25.

3. Rita S. Gottesman, *The Arts and Crafts in New York, 1726–1776* (New York: New-York Historical Society, 1938), pp. 356–358.

4. U.S. Bureau of the Census, *Historical Statistics of the United States, Colonial Times to 1970*, vol. 2 (Washington, D.C., 1975), pp. 1176–1177.

5. Timothy Pitkin, *A Statistical View of the Commerce of the United States of America* (1835; reprint ed., New York: Johnson, 1967), pp. 586–587.

6. George Dodd, *Days at the Factories* (1843; reprint ed., New York: Augustus M. Kelley, 1967), p. 432.

7. Franklin Bowditch Dexter, ed., *New Haven Town Records, 1649–1662* (New Haven: New Haven Colony Historical Society, 1917), p. 116.

8. *American State Papers. Documents, Legislative and Executive, of the Congress of the United States . . . Commencing March 3, 1789, and Ending March 3, 1815*, vol. 6: *Finance*, vol. 2 (Washington, D.C.: Gales and Seaton, 1832), p. 674.

9. *American State Papers*, vol. 6: *Finance*, vol. 2, p. 426.

VIII / COOPER

1. William Bradford, *Of Plymouth Plantation, 1620–1647*, ed. Samuel Eliot Morison (New York: Alfred A. Knopf, 1952), pp. 140, 163, 176.

2. John Winthrop, *Papers*, vol. 3, 1631–1637 (Boston: Massachusetts Historical Society, 1943), pp. 41–44.

3. Nathaniel B. Shurtleff, ed., *Records of the Governor and Company of the Massachusetts Bay in New England*, vol. 2, 1642–1649 (Boston: William White, 1853), pp. 250–251.

4. *Public Records of the Colony of Connecticut, Prior to the Union with the New Haven Colony, May, 1665* (Hartford: Brown & Parsons, 1850), pp. 515–516.

5. *Public Records of the Colony of Connecticut . . . May, 1665*, pp. 67–68.

6. *Public Records of the Colony of Connecticut . . . May, 1665*, p. 200.

7. Tench Coxe, *A View of the United States of America, in a Series of Papers Written at Various Times Between the Years 1787 and 1794* (Philadelphia: William Hall, 1794), pp. 419, 475.

8. Kenneth Kilby, *The Cooper and His Trade* (London: John Baker, 1971), p. 171.

IX / TOOLS TO MEASURE AND HOLD

1. Richard More, *The Carpenters Rule* (1602; reprint ed., New York: Da Capo Press, 1970).

2. Wilhelm Bedwell, *Mesolabium Architectonicum* (1631; reprint ed., New York: Da Capo Press, 1970), table.

3. *Public Records of the Colony of Connecticut, Prior to the Union with the New Haven Colony, May, 1665* (Hartford: Brown & Parsons, 1850), pp. 85, 159–160.

4. Rita S. Gottesman, *The Arts and Crafts in New York, 1726–1776* (New York: New-York Historical Society, 1938), p. 307.

5. A. Viires, *Woodworking in Estonia* (Jerusalem: Israel Program for Scientific Translations, 1969), pp. 60–62.

X / DECLINE OF THE HAND TOOL

1. Quoted in Charles E. Peterson, "Early Lumbering: A Pictorial Essay," in *America's Wooden Age: Aspects of its Early Technology*, ed. Brooke Hindle (Tarrytown, N.Y.: Sleepy Hollow Restorations, 1975), p. 74.

2. J. Leander Bishop, *A History of American Manufactures from 1608 to 1860*, 3rd ed., vol. 2 (1868; reprint ed., New York: Johnson Reprint Corp., 1967), p. 332.

3. [John K. Porter], *Argument Against the Extension of the Woodworth Patent* (n.p., n.d.), p. 18.

4. Henry Hall, "Report on the Ship-Building Industry of the United States," p. 142, in *Tenth Census of the United States*, vol. 8, *Special Reports* (Washington, D.C., 1884).

5. William B. Hilton, "A Checklist of Boston Plane Makers," *Chronicle of the Early American Industries Association* 27 (1974): 23–27.

6. Elliot M. Sayward and William M. Streeter, "Planemaking in the Valley of the Connecticut River and the Hills of Western Massachusetts," *Chronicle of the Early American Industries Association* 28 (1975): 21–29.

7. Kenneth D. Roberts, *Wooden Planes in 19th Century America* (Fitzwilliam, N.H.: Ken Roberts, 1975), pp. 189–198.

BIBLIOGRAPHY

The following bibliography is a selection from the works consulted. Many titles cited in the preceding References are not repeated. The bibliography is in three parts: five basic books on tools; a list of works consulted; and a list of representative trade catalogs that are illustrative of the tools of woodworking trades.

SOME BASIC BOOKS

Goodman, W. L. *The History of Woodworking Tools.* London: G. Bell and Sons, 1964. 208 pp.

A thorough work by one of the most knowledgeable writers in the field. Well documented and illustrated. The chief emphasis is on tools of England and continental Europe.

Hummel, Charles F. *With Hammer in Hand; the Dominy Craftsmen of East Hampton, New York.* Charlottesville: University Press of Virginia, 1968. 424 pp.

Systematic study of the tools of the Dominy family, now in the Henry Francis du Pont Winterthur Museum. The Dominys were primarily cabinetmakers and clockmakers. The tools, many of which they made for their own use and are limited to those of their specific crafts, are described in detail with accompanying illustrations. Similar treatment is given to Dominy furniture and clocks.

Mercer, Henry C. *Ancient Carpenters' Tools.* 5th ed. New York: Published for the Bucks County Historical Society by Horizon Press, 1975. 331 pp.

The first substantive book on American woodworking tools, by the pioneer in their collection, preservation, and identification, and still an excellent source of detailed information. The author describes the tools which were in his vast collection, occasionally notes their provenance, but rarely gives information about their makers. First published in 1929.

Salaman, R. A. *Dictionary of Tools Used in the Wood-working and Allied Trades, c. 1700–1970.* New York: Charles Scribner's Sons, 1975. 545 pp.

A comprehensive work, researched in depth and carefully written. It is an essential reference work for the identification and understanding of thousands of varieties of hand tools. Fully illustrated with original drawings and cuts from contemporary tool catalogs.

Sloane, Eric. *A Museum of Early American Tools.* New York: Wilfred Funk, 1964. 108 pp.

Drawings by the author-artist which capture the flavor of the early woodworking crafts, in a fine introductory book. Brief explanatory text accompanies the illustrations of tools and farm implements.

WORKS CONSULTED

America's Wooden Age: Aspects of its Early Technology. Edited by Brooke Hindle. Tarrytown, N.Y.: Sleepy Hollow Restorations, 1975.

Arnold, James. *The Shell Book of Country Crafts.* London: John Baker, 1968.

Bishop, J. Leander. *A History of American Manufactures from 1608 to 1860.* 3rd ed., rev. and enl. 3 vols. 1868. Reprint. New York: Johnson Reprint Corp., 1967.

Bridenbaugh, Carl. *The Colonial Craftsman.* New York: New York University Press, 1950.

Briggs, Martin S. *The Homes of the Pilgrim Fathers in England and America (1620–1685).* London: Oxford University Press, 1932.

Carpenters' Company of the City and County of Philadelphia. *The Rules of Work, 1786.* Reprint. Annotated by Charles E. Peterson. Princeton, N.J.: Pyne Press, 1971.

Chronicle of the Early American Industries Association. Vol. 1, 1933 to date.

Clark, Victor S. *History of Manufactures in the United States.* Rev. ed. 3 vols. New York: McGraw-Hill, 1929.

Diderot, Denis. *Encyclopédie, ou Dictionnaire Raisonné des Sciences, des Arts et des Métiers.* 17 vols. Paris: Briasson [etc.], 1751–1765.

———. *Recueil de Planches.* 11 vols. Paris: Briasson [etc.], 1762–1772.

A Directory of Sheffield, Including the Manufacturers of the Adjacent Villages. 1787. Reprint. New York: Da Capo Press, 1969.

Henry Disston & Sons, Inc. *The Saw in History.* 8th ed. Philadelphia, 1925.

Felton, William. *A Treatise on Carriages.* London: The Author, 1796.

Goodman, W. L. *British Plane Makers from 1700.* London: G. Bell & Sons, 1968.

Hall, Henry. *Report on the Ship-Building Industry of the United States.* Tenth Census of the United States, vol. 8. Washington, D.C.: Government Printing Office, 1884.

Hazen, Edward. *Popular Technology; or, Professions and Trades.* 2 vols. New York: Harper & Brothers, 1855.

Hegel, Richard. *Carriages from New Haven.* Hamden, Conn.: Archon Books, 1974.

Hummel, Charles F. "English Tools in America; the Evidence of the Dominys," *Winterthur Portfolio* 2 (1965): 27–46.

Isham, Norman Morrison. *Early American Houses, and Glossary of Colonial Architectural Terms.* 1928. Reprint. New York: Da Capo Press, 1967.

Isham, Norman M., and Brown, Albert F. *Early Connecticut Houses; an Historical and Architectural Study.* New York: Dover Publications, 1965.

Jenkins, J. Geraint. *Traditional Country Craftsmen.* New York: Frederick A. Praeger, 1967.

Jones, P. d'A., and Simons, E. N. *Story of the Saw.* Manchester, Eng.: Newman Neame (Northern) for Spear and Jackson, Aetna Works, Sheffield, 1961.

Kauffman, Henry J. *American Axes; a Survey of Their Development and Their Makers.* Brattleboro: Stephen Greene Press, 1972.

Kelly, J. Frederick. *The Early Domestic Architecture of Connecticut.* 1924. Reprint. New York: Dover Publications, 1963.

Kilby, Kenneth. *The Cooper and His Trade.* London: John Baker, 1971.

Knight, Edward H. *American Mechanical Dictionary.* 3 vols. Boston: Houghton, Mifflin, 1876.

———. *New Mechanical Dictionary.* Boston: Houghton, Mifflin, 1883.

Martin, Thomas. *The Circle of the Mechanical Arts.* 2nd ed. London: J. Bumpus, 1818.

Moxon, Joseph. *Mechanick Exercises, or, The Doctrine of Handy-Works.* 14 pts. London: Joseph Moxon, 1677–1680.

Nicholson, Peter. *New Carpenter's Guide.* Improved and enlarged ed. London: Jones, 1826.

Nicholson File Company. *A Treatise on Files and Rasps.* Providence, R. I., 1878.

Oakley, Kenneth P. *Man the Tool-Maker.* Chicago: University of Chicago Press, 1957.

Petrie, W. M. Flinders. *Tools and Weapons.* London: British School of Archaeology in Egypt, 1917.

Roberts, Kenneth D. *Wooden Planes in 19th Century America.* Fitzwilliam, N.H.: Ken Roberts, 1975.

Roberts, Kenneth D., and Roberts, Jane W. *Planemakers and Other Edge Tool Enterprises in New York State in the Nineteenth Century.* Cooperstown, N.Y.: New York State Historical Association, 1971.

Roubo, André Jacob. *L'Art du Menuisier.* 4 vols. 1769–1775. Reprint. 4 vols. in 3. Paris: Léonce Laget, 1976.

Sellens, Alvin. *The Stanley Plane; a History and Descriptive Inventory.* Augusta, Kan.: Alvin Sellens, 1977.

Smith, Joseph. *Explanation or Key, to the Various Manufactories of Sheffield.* 1816. Reprint. South Burlington, Vt.: Early American Industries Association, 1975.

Story, Dana A. *The Building of a Wooden Ship: "Sawn Frames and Trunnel Fastened."* Barre, Mass.: Barre Publishers, 1971.

Sturt, George. *The Wheelwright's Shop.* 1923. Reprint. Cambridge, Eng.: University Press, 1963.

Tarr, László. *The History of the Carriage.* Trans. by Elisabeth Hoch. New York: ARCO Publishing Co., 1969.

Viires, A. *Woodworking in Estonia.* Trans. from Estonian. Jerusalem: Israel Program for Scientific Translations, 1969.

Welsh, Peter C. "Woodworking Tools, 1600–1900," *Contributions from the Museum of History and Technology, Paper 51. United States National Museum Bulletin* 241: 179–227. Washington, 1965.

Wildung, Frank H. *Woodworking Tools at Shelburne Museum.* Shelburne, Vt.: Shelburne Museum, 1957.

Winthrop, John. *Papers.* 5 vols. Boston: Massachusetts Historical Society, 1929–1947.

REPRESENTATIVE TRADE CATALOGS

A. F. Brombacher & Co. *Tools for Coopers and Gaugers, Produce Triers. Catalogue no. 20.* 1922. Reprint. N.p., n.d.

Buck Bros. *Price List of Chisels, Plane Irons, Gouges . . .* *&c.* 1890. Reprint. Fitzwilliam, N.H.: Kenneth Roberts, 1976.

W. H. Carr & Co. *American Manufactured Hardware, &c.* 1838. Reprint. N.p.: Early American Industries Association, 1972.

Collins & Co. *Illustrated Catalogue.* 1921. Reprint. Levittown, N.Y.: Early Trades & Crafts Society, 1974.

Hammacher, Schlemmer & Co. *Tools for All Trades, and Supplies. Catalog No. 355.* New York, [1908?].

Hart, Bliven & Mead Manufacturing Co. *Catalogue and Price List.* New York, 1873.

Mack & Co. *Price List of Genuine "D. R. Barton" Planes, Edge Tools, &c.* Rochester, N.Y., 1909.

Nicholson File Company. *Illustrations of Files, Rasps and Tools.* Providence, R.I., 1894.

Ohio Tool Company. *Illustrated Price List No. 23.* Columbus, [1910?].

Joshua Oldham, firm. *Catalogue and Price List. To Which is Appended a Treatise on Saws, Their History, Manufacture, and Use.* 1887. Reprint. South Burlington, Vt.: Early American Industries Association, 1976.

Peck, Stow & Wilcox Co. *Illustrated Catalogue and Price List.* New York, 1898.

Stanley Rule and Level Company. [*Catalogues* and *Price Lists* of rules, levels, and carpenters' tools, 1855–1909, in various reprint editions.] V.p., v.d.

L. & I. J. White Co. *Catalogue of Coopers' Tools, Including Turpentine Tools.* 1912. Reprint. N.p., n.d.

INDEX

Page numbers in boldface type refer to black-and-white illustrations